More Praise for *How to Win Client Business When You Don't Know Where to Start*

"*How to Win Client Business When You Don't Know Where to Start* skillfully captures the fundamental and critical insights about how to really 'sell' professional services. Doug Fletcher synthesizes years of valuable experience into a pragmatic dialogue that is pure gold for any professional working to develop clients and business. Whether you have never sold a dollar, or you have years of client development experience, you will find these invaluable lessons instructive, refreshing and inspiring."

—Walt Shill, Global Commercial Officer, ERM

"If you are highly skilled at doing client work—but are puzzled by how to 'get' the client work, Doug's wonderful book demystifies the process for us. This book contains proven techniques that will make you more credible, trusted and in demand. Read this today, apply the lessons and watch your practice grow."

—David T. Richardson, President & CEO, Cognision

"A rare and terrific resource for lawyers, consultants, and other providers of professional services, who are often intimidated at the concept of selling their time and expertise. Winning client business is different from selling widgets. Doug provides real world examples and pragmatic, doable strategies that even the most sales-adverse among us can use to build a successful business."

—Judy Selby, Partner, Hinshaw & Culbertson LLP

HOW TO WIN CLIENT BUSINESS

WHEN YOU DON'T KNOW
WHERE TO START

A Rainmaking
Guide
for Consulting
and Professional
Services

DOUG FLETCHER

WILEY

Published by John Wiley & Sons, Inc., Hoboken, New Jersey.
Published simultaneously in Canada.

For general information on our other products and services or for technical support, please contact
our Customer Care Department within the United States at (800) 762-2974, outside the United States
at (317) 572-3993 or fax (317) 572-4002.

Wiley publishes in a variety of print and electronic formats and by print-on-demand. Some material
included with standard print versions of this book may not be included in e-books or in print-on-
demand. If this book refers to media such as a CD or DVD that is not included in the version you
purchased, you may download this material at http://booksupport.wiley.com. For more information
about Wiley products, visit www.wiley.com.

Library of Congress Cataloging-in-Publication Data

Names: Fletcher, Doug, author.
Title: How to win client business when you don't know where to start : a
  rainmaking guide for consulting and professional services / Doug
  Fletcher.
Description: Hoboken, New Jersey : John Wiley & Sons, Inc., [2021] |
  Includes bibliographical references and index.
Identifiers: LCCN 2021005019 (print) | LCCN 2021005020 (ebook) | ISBN
  9781119676904 (cloth) | ISBN 9781119676966 (adobe pdf) | ISBN
  9781119676928 (epub)
Subjects: LCSH: Consulting firms. | Consultants. | Consulting
  firms—Marketing. | Consultants—Marketing.
Classification: LCC HD69.C6 .F548 2021 (print) | LCC HD69.C6 (ebook) |
  DDC 001—dc23
LC record available at https://lccn.loc.gov/2021005019
LC ebook record available at https://lccn.loc.gov/2021005020

Cover Design: Wiley

SKY10029798_091021

To Duncan and Abby,
who inspire me every day and give me so much hope for the future.
I love you both and am proud to be your dad.

Contents

THE RAINMAKER'S JOURNEY

Introduction

Selling a Service Is Different (and Harder) Than Selling a Product

Here's a simple thought experiment for you to ponder:

Why is it that we have no problem whatsoever buying a home after an hour-long walk-through, yet will agonize for months in choosing an architect if building a custom home?

We're comfortable spending $500,000 on a house after a short visit, but require months to choose an architect who will cost us a small fraction of this amount. Alternatively, we'll spend $100,000 on a new Tesla Model S convertible after a 30-minute test drive, but struggle for months to choose a financial planner.

Hmmm, why is this? Selling a service is much different – and harder – than selling a product. Or, more appropriately, buying a service is a much different experience than buying a product. Something interesting is going on here that I think warrants further thought and discussion.

OK, maybe you've never pondered this before. But I have. I'll admit, among friends at least, that I spend many of my waking hours thinking about things like this: *how clients buy* and *how to win client business*. These are fascinating topics, and oddly, not studied nearly enough.

If you, too, find this question interesting, maybe you're a fellow journeyman in this quest to better understand what goes on inside the heads of prospective clients and how to win more client business. Let's join together and travel in search of answers to questions such as these.

If you aspire to become a partner in your firm, I'm sure these questions resonate with you. Maybe you're a consultant, designer, accountant, engineer, or financial advisor; or an attorney, investment banker, or web developer. Or maybe you're considering going out on your own as a solo practitioner or starting up a new firm with a few colleagues to provide your expertise to the world. Regardless of who you are, you likely want to build better relationships with your clients. I want that too. That's why I'm writing this book for you.

The truth is that we have never been taught how to win client business. We're taught to do accounting, practice law, invest money, design bridges, and create websites. But rarely, if ever, do we receive any training when it comes to selling the expert services that we provide. It's a sad truth and it limits our ability to have a successful and fulfilling career. My aspiration for this book is to help remove much of the mystery of how to win client business and to accelerate your career success.

Twenty years ago, I left the world of large organizations and co-founded a management consulting firm with two of my work colleagues. Our clients were among the Fortune 500. I was in my early 30s at the time, and very sure of myself. Frankly, despite my first-rate work experiences and education, I did not have much exposure to selling.

What I didn't appreciate at the time was how hard it can be to win client business, especially when you're just getting started. Life has a way of teaching us some humbling lessons at times, and my self-confidence was quickly knocked down a few notches once I discovered I knew next to nothing about how clients buy.

Of all the aspects of growing a firm, I find the topic of winning client business to be the most interesting and challenging. It's not to say that the other aspects of business are less important. Being an inspiring leader, creating a positive work culture, and implementing vital business processes – yes, all of these are crucial to succeeding in any business.

I am curious about understanding the client's buying decision journey and how we can influence it. It's my passion. Strike that; it borders on an obsession. It's what I wake up at 2:00 a.m. thinking about. It's what gets me up in the morning excited to start the day. And my goal is to share with you what I've learned in the hope that it will help you reach your career aspirations.

Despite more than a few missteps on my part, my partners and I managed to win more than our fair share of business over the next 15 years. Since selling our firm in 2014, I have taken a deeper dive into the study of business development for consulting and the professional services.

The term *business development* is a euphemism used to describe sales and marketing activities in professional services. For a variety of historical and cultural reasons, we don't use the word *sales* or *selling* in professional services. Throughout the book I use the term *client development* instead of business development. In this shift, I believe it places the proper emphasis on the client. As we'll soon learn, our success begins with an understanding of how clients think and make buying decisions.

Our investigation into the client's mind will include a mix of interrelated fields: human psychology, consumer behavior, marketing strategy, and behavioral economics, and maybe a wee bit of philosophy and history mixed in for good measure.

In addition to lessons learned throughout my career, you'll hear advice from many successful professionals. I have benefited from the wisdom shared by many rainmakers from every imaginable profession.

If I had known at 30 what I know now, the arc of my career could have been vastly different. So join me, if you will, on this quest to better understand the client's buying decision journey and improve our ability at winning client business. If we can do this, together, I'm confident we'll have more successful and satisfying careers.

If I'm So Smart,
Why Do I Feel So Stupid
about Selling?

CHAPTER

1

Things Rainmakers Do That Most of Us Don't: The Five Rainmaker Skills

Universities Don't Teach Us and Our Firms Don't Train Us

Every professional firm needs more people who develop new business. Accountants, actuaries, attorneys, engineers and management consultants are all familiar with this problem. Bright, young, technical talent is always available. Seasoned project managers usually are. But never are there enough rainmakers.
—Ford Harding, author, *Creating Rainmakers*

If you're old enough to remember John Grisham's 1995 novel *The Rainmaker*, you understand what we mean by the term *rainmaker*. Grisham's novel was a huge hit, rapidly stepping into the number one spot on the *New York Times* best-seller list. According to Grisham's publisher, Doubleday, it was the fastest-selling hardcover book ever at the time. Francis Ford Coppola's movie that followed a couple of years

7

later starring Matt Damon was a box-office hit. If you were born after, say 1985, you may be thinking to yourself, "I have no idea what we're talking about here."

What Is a Rainmaker?

What is a rainmaker? A rainmaker generally refers to a partner in a professional services firm who is skilled at bringing in client business. Rainmakers:

- Generate leads for new business
- Turn leads into new clients
- Are skilled at turning existing clients into referrals and repeat business
- Keep many people in their firms employed
- Are highly respected and frequently have a lot of influence in their firms

According to University of Wisconsin professors Marc Galanter and Thomas Palay, the first appearance of the term "rainmaker" can be traced to the 1970s. Before that, we simply referred to rainmakers as *business-getters*. Rainmakers are business-getters.

If you want to become a partner in your firm, or to succeed in your own practice, it's hard to succeed without becoming a rainmaker.

Why Do I Feel So Stupid About Selling?

If you've been at your profession for long, or you've already hung out your own shingle, it's no secret that we have to win client business if we are to become successful. Of that, I'm confident – as I am in the laws of gravity. You won't make partner if you can't make the cash register ring. Certainly you won't stay in business long as a solo practitioner if you don't have enough clients.

It's funny, in a sad sort of way, that they don't teach us how to do this at law school, business school, engineering school, or architecture school. We spend years of our young lives and huge sums of money learning our professions. But, ironically, we're never taught the one thing that our future success depends upon.

I should know. I teach at a college of business. Nowhere in our curriculum is there any course that would provide a young professional with the knowledge and skills to win client business in the professional services. And my college is no different than any other business school out there.

Furthermore, I graduated from a top MBA program. How to win client business isn't taught there, either. Nor is it taught at any other top MBA program. I also graduated from an excellent engineering program – again, never discussed. The one thing that could have a huge impact on the success of our professional career is not spoken of.

It's akin to Lord Voldemort in *Harry Potter* – that which must not be named. I won't go into why we aren't taught this in school. That's a rabbit hole I'll save for another day. Suffice it to say you've never been taught how to win clients at the university.

Nor are we taught how to do this by our firms. A few do provide a seminar here or there, but our firms don't provide any structured, systemic training to high-potential, pre-partner staff on how to win client business. The approach taken by most firms is simply to throw everyone into the deep end of the pool and wait to see who dog-paddles their way out of it. This is true of every profession that I have witnessed. The downside to the sink or swim approach is that we lose a lot of highly talented people along the way. The attrition rate at the partner juncture is high. I guess it's at least 50%.

Pete Sackleh, who previously held the positions of managing director of Deloitte University and executive director of KPMG's Learning & Innovation Center, knows something about the transition to partner level. He had a front-row seat at two of the most successful professional services firms in the world.

When I spoke with Pete recently about the high failure rate at the partner juncture, he felt my 50% estimate was too low: "The failure rate at reaching the partner levels is much higher than 50%. I would guess it's closer to 70%. Look at the org chart of these large professional services firms and do the math for yourself."

Partner track attrition is caused by one of two root causes:

1. Failure at winning client business
2. Self-selecting out

With better training, there would be a higher success rate and a lower attrition rate. You would think professional firms would provide more training on client development. The truth is that most of those in charge of training young professionals don't know where to go for help. Frankly, there really isn't that much out there to choose from. Most of the sales training offered is geared toward products. And, as we'll discuss, selling a service is different than selling a product.

Furthermore, most of the partners that are successful rainmakers have a difficult time explaining what it is they do. It's the reason why most successful people, in any given field, are often not the best teachers. Try asking Michael Jordan for tips on how to play basketball, or Serena Williams for suggestions on how to improve your tennis swing. It's very hard for them to articulate why they are so good at what they do. The best coaches are rarely the best players at their sport.

For decades, Vic Braden was the go-to coach for many of the world's best tennis players. Vic always made a point of asking his players why and how they play the way they do. "Out of all the research that we've done with top players, we haven't found a single player who is consistent in knowing and explaining exactly what he or she does," says Braden. Similarly, we're not going to get much helpful coaching from the best rainmakers in our firms.

Another area sorely lacking is in the limited number of practical books available on the topic. Relative to all other business topics, there just isn't much out their written specifically on client development in consulting and professional services. Want to learn about leadership? You could fill London's Wembley Stadium with the books available from Amazon. Want to find a good book on becoming a rainmaker? Good luck. You can count on one hand the number of good books published on the topic over the past decade. There are a few very good books – for example, those by David Maister and Ford Harding. But many of these classics are over 20 years old. So, even if a professional were looking for self-help in this area, there isn't much to grab on to.

So, I would say to you, cut yourself some slack. If you're feeling stupid about selling your services, please stop beating yourself up. Why feel bad about not knowing something that you have never been taught? You shouldn't. My hope in writing this book is that it gives highly talented professionals a place to start in learning the craft of winning client business.

The Five Rainmaker Skills, the Focus of This Book

If I could go back in time to have a beer with my younger self, these five skills are things that I wish I had known. In my experience and observations over the past 25 years, these are the important skills that successful rainmakers practice that differentiate themselves from the rest of the pack to have long-term success at consistently winning new client business.

I say "consistently" because we can all get lucky from time to time. Winning one new client can be attributed to good fortune – being in the right place at the right time or being on the receiving end of a highly desperate situation. But long-term, we don't want to leave our success to chance. Therefore, we need to learn to practice these rainmaker skills.

The good news is these skills aren't rocket science. While they may seem at first like a sorcerer's bag of tricks, they are not magic. Rather, they are practical skills based upon sound principles of consumer behavior and marketing strategy with a healthy dose of practice, hard work, and trial and error. Let's unpack these five skills at a high level. (See Figure 1.1.) We'll dive into them in much more detail in future chapters.

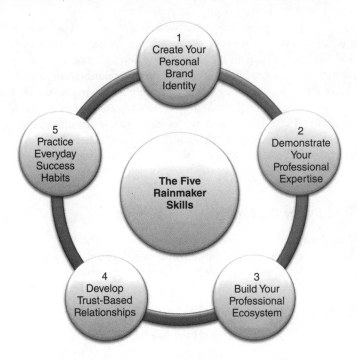

FIGURE 1.1   The Five Rainmaker Skills

Skill 1: Create Your Personal Brand Identity

Establishing our personal brand identity begins with deciding a) what we want to be known for and b) who we wish to serve. These two topics, our chosen field of expertise and our target audience, are the sharp end of the client development spear. We'll talk about the importance of focus and how, paradoxically, that sets us up for success.

Interestingly, the more we focus, the more we are seen as experts, and this leads to greater brand identity and respect. We can be known for being good at anything, but we can't be known for being good at everything. Until we choose and effectively communicate what makes us unique, we'll have difficulty in consistently winning client business.

Skill 2: Demonstrate Your Professional Expertise

Clients need clues that we are really good at what we do. Remember the thought experiment from the book's introduction about buying the house versus hiring the architect? The reason it's easier to buy a house is that it is much easier to assess the quality of a "thing" than determine how happy we will be with a "service." A product is used. A service is experienced. It's the difference between *what is* and what *could be*.

In order for a prospective client to have confidence they will be satisfied with our service, we have to provide evidence that we have done great work for similar people or organizations in the past. I often refer to these as channel markers, in the way that boats need channel markers to navigate safely from sea to port without running aground. We must demonstrate our expertise in specific ways that helps a prospective client feel comfortable in choosing to work with us.

Skill 3: Build Your Professional Ecosystem

It is often said in professional services that we are in the relationship business. Clients hire people with whom they have a relationship. Or, in the absence of knowing someone who can assist them, they rely heavily on the advice of others. Our ability to win client business is based upon the quality of our professional network.

I prefer to use the term "ecosystem" rather than "network." Ecosystem is a better word, I believe, because networking has taken on a negative tone for some. Ecosystems are a complex mesh of mutually supportive, beneficial relationships. Our professional ecosystem is comprised of the people with whom we have relationships. The most successful rainmakers spend considerable time building their professional ecosystem. Not in a superficial way (say, LinkedIn connections you've never met), but in ways that are genuine and real.

Skill 4: Develop Trust-Based Relationships

Professional ecosystems flourish when the ties between its members are deepened by trust-based relationships. This is the second phase of ecosystem building. Once established, relationships are strengthened by a commitment to helping one another succeed. Successful rainmakers value the people in their professional network, and dedicate time to helping these individuals succeed.

Real relationships are the key to winning client business, not a large advertising budget or digital marketing savvy. Helping others is at the heart of all human relationships, personal and professional. Real relationships are built upon respect and trust. Real relationships are built over time by being honest, helpful, and caring. We'll learn from the relationship habits of the most successful rainmakers so that we can apply them to our valued partners in our professional ecosystem.

Skill 5: Practice Everyday Success Habits

Success at consistently winning client business comes from dedicating time every day to practicing the rainmaker skills. Practicing the rainmaker skills is not a sometimes thing; it's something that successful rainmakers practice every day of the year. We'll learn that success comes from building a personal client development system that works for you.

By tailoring your approach to your strengths and preferences, you'll be more willing to stick with it. No two rainmakers' systems are identical, but they are alike in that they are practiced consistently. Developing good daily rainmaker habits early in one's career creates momentum that will become the foundation of future success.

Moving Forward with the Five Rainmaker Skills

In this book we'll learn the five rainmaker skills and improve our success at winning new client business. We'll examine approaches that others have found success with. And you'll be able to pick and choose from these examples to find an approach that best fits your interests and strengths.

But I'm getting ahead of myself. Let's step back and first take a closer look at the client's buying decision journey. Understanding *how clients buy* is the consumer behavior piece of this rainmaker puzzle. Rainmakers have a keen sense of how clients think and what they need to feel comfortable in choosing to work with us. Understanding the clients buying decision journey is the first step in becoming more successful as a rainmaker.

The Importance of Doing Great Work and Repeat Clients

It is hard to have a meaningful discussion about winning client business without discussing two very important topics:

1. **The Importance of Doing Great Work**
2. **The Value of Repeat Clients**

Doing great work is vital to winning client business. The two can't be separated. If we don't do good work, we just won't be around for long.

But this is not a book on *how to do great work*; this is a book about *how to get the work*. My focus is not meant in any way to diminish the importance of being really good at your craft. It's just that most of us are already skilled at doing great work. Universities, professional organizations, continuing education, and firm training all contribute to the ongoing emphasis on how to do our jobs well.

There's a distinct difference, though, between the doin' and the gettin'. Frankly, it's harder to find professionals who are skilled at getting the work than doing the work; hence the high-attrition rate at the partner juncture and the relative scarcity of successful solo-practitioners.

The second topic, the value of repeat clients, is also incredibly important. This was the case in my consulting firm. When we sold North Star Consulting Group in 2014, we had a considerable client base that had been with us since we opened our doors in 1999. Without these great, long-term clients, our firm would not have been successful.

In many professions – if not most – consistent, repeat business from existing clients is the backbone of the firm. For many professionals, repeat clients represent 80% of their business. Without repeat business, the time commitment of finding new clients would consume nearly all of our week and there wouldn't be time for much else.

It's been well-documented that repeat clients are in many cases the most profitable. Over time, loyal clients often base their decision to work with us less on cost-sensitivity and more on the quality of the service and relationship. And there is less of a learning curve on our end in serving them well. Additionally, without doing great work for our existing clients, referrals to new client business would be few and far between.

Although these two topics are important, they are not the focus of this book. My passion lies in how to win *new* business, clients we have never served before. Not because it is more important than doing great work, but because winning new clients is the primary criterion of those on the partner track and the newly self-employed. And, if we've already made partner, new client business will bring growth and vibrancy to our baseline practice. Without learning to generate new client work, all businesses will eventually plateau or decline.

References

John Grisham. *What is a rainmaker? The Rainmaker*. Doubleday, 1995.

Mary Tabor. "Nobody Can Keep Pace with John Grisham." *New York Times*, *April* 19, 1995.

Ford Harding. *Creating Rainmakers: The Manager's Guide to Training Professionals to Attract New Clients*. John Wiley & Sons, 2006.

David Maister et al. *The Trusted Advisor*. Touchstone, 2001.

Pete Sackleh quote: Interview by *How to Win Client Business* research team, 2020.

Vic Braden story: Malcolm Gladwell. *Blink: The Power of Thinking Without Thinking*. Little, Brown and Company, 2005.

2

How Clients Buy

Understanding the Client's Buying Decision Journey

> It is particularly hard for people to make good decisions when they have trouble translating the choices they face into the experiences they will have.
>   —Richard H Thaler, University of Chicago, Nobel Laureate in Economics

I was having dinner recently with my good friend John Senaldi, who lives in the Bay Area. As we enjoyed a glass of 12-year-old Talisker, it occurred to us that it had been 30 years since we worked together at GE Aerospace in Syracuse, New York. We were in our early 20s at the time.

After catching up on our kids and other small talk, our conversation drifted to our professional lives. John had spent most of his career in medical technology. By any measure, he had a very successful career leading to the CEO position at a tech company he helped turn around and sell.

John had recently transitioned from corporate life to the role of executive coach and strategic advisor. He now leveraged his considerable experience in helping CEOs navigate through difficult business issues.

"How's your new consulting practice going?" I asked.

"Well, I gotta admit, it's been a bit of a transition to the new role as coach and advisor," John replied. "The hardest part for me is learning how to market my services to prospective clients."

"Tell me about your successes this past year," I prompted.

John replied, "It took a while, but I have managed to land two really good clients. I'm helping a couple of CEOs with issues I'm familiar with. I'm having a great time, and I feel like I'm making a difference."

"Where'd the clients come from?" I asked.

"Well," John paused, "my first two clients were people I had known for a while…. I worked with them previously. They knew me and felt that I could help them grow their businesses."

I smiled to myself, knowing John's story well. Many professionals I speak with have similar stories of how they got their first clients. I always make a point of asking rainmakers where their most recent clients came from. No two client stories are identical, but they have some interesting things in common.

If we are to become successful rainmakers, we have to build authentic relationships with those we wish to serve. Our clients have to know us, they have to respect our professional abilities, and they have to trust that we are honest. And, in the absence of this, we have to be strongly recommended by someone the prospective client respects and trusts.

If this is true, what do we have to do as professionals to get to know people, and have others respect and trust us? There's a lot to unpack there. And, we'll get to that soon, but the first step is to learn to think like a client. We have to understand the client's buying decision journey. The five rainmaker skills that we'll learn are built upon the foundation of the client's buying decision journey. In understanding how clients buy, we'll begin to see why these rainmaker skills work.

The Seven Elements of the Client's Buying Decision Journey

The above description of how clients buy is the *CliffsNotes* version. It's true that clients have to know, respect, and trust us, but there is much more to it. Tom McMakin and I spent two years examining the longer version of the client's journey. We covered this in depth in our book *How Clients Buy: A Practical Guide to Business Development for Consulting and Professional Services*.

In *How Clients Buy*, we outline the seven elements of the clients buying decision journey. (See Figure 2.1.)

Our framework of the client's decision-making process comes from our own personal experiences, in addition to interviews with successful rainmakers in a wide range of professions. Following are the highlights of what we learned.

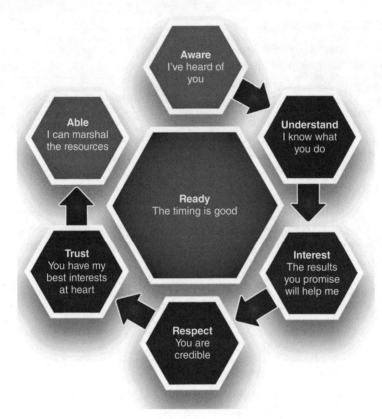

FIGURE 2.1   The Seven Elements of the Client's Buying Decision Journey

Element 1: Awareness

A prospective client becomes *aware* of you. This could be someone you worked with years ago, someone you met at a conference, someone who read an article you wrote in a trade journal, or perhaps someone introduced to you at a friend's holiday party. Whatever the case, your existence is known to a prospective client.

Element 2: Understanding

Once your existence is known, the prospective client needs to clearly *understand* what you do, who you serve, and how you are unique. It's not enough to know that you exist. In order for others to hire you or recommend you to others, they need to be able to clearly articulate your area of expertise, who you serve, and what differentiates you from other service providers.

Element 3: Interest

Prospective clients have to be *interested* in what you do. They have to see that what you do could help them or, alternatively, help others they know. Interest lies at the intersection of your expertise and the problems of the world. Until someone is interested in the work that you do, they are simply friends, colleagues, or acquaintances. Your existence and expertise are filed away, but the client's buying decision journey is far from over.

Element 4: Respect

Respect relates to your professional credibility. A prospective client, or someone who could potentially recommend you, has to believe that you are really good at what you do. Naturally, we wouldn't want to hire someone who wasn't good at what they do, or recommend someone to a friend if we didn't believe they were highly skilled. Prospective clients and colleagues may struggle with this because it's often hard to tell if we are really good at our profession. They look for clues – called credibility markers – to help them gauge our level of professional competence.

Element 5: Trust

Others must believe that you are a *trustworthy* individual if they would ever consider hiring you or recommending you to a friend. Are you honest? Are you ethical? Are you the kind of person that looks out for others' interests, or are you self-serving? Trust is the glue that holds the world together. It's hard for others to see trust in our heart. Our trust is earned over time by doing things that demonstrate our character.

Element 6: Ability

Ability refers to a prospective client having the budget, decision authority, and organizational support to hire you. Without funding, we won't be in a position to help an individual or organization, unless we're doing pro bono work for a worthy cause. Additionally, when our prospective client is a part of an organization, does the person have the decision authority and support from her team? Until our prospective client has the budget, decision authority, and organizational buy-in, we're a ways away from winning their business.

Element 7: Readiness

As with many things in life, timing is everything. This is often the case with winning client business. Until your project or service has risen to the top of their priority list, the client isn't *ready* to buy. Sometimes patience is the fastest way to get what you want. Being attuned to the client's priorities will help us better understand when it's the right time for us to help others.

Putting the Seven Elements Framework to Use

Interestingly, we found from our research that these seven elements of the client's buying decision journey are not necessarily linear. They could, in theory, happen in the exact order as we have them outlined. A prospective client could become aware of us, learn what we do, have an interest in our services, grow to respect our professional ability, trust us, have the funds, and be ready to get started. But that is not often the way in which things play out.

A new client could have met us ten years ago at a conference, grown to respect our professional expertise over time, and developed a belief that we are trustworthy, but didn't have an interest in working with us at the time. It may be years later before they find themselves searching for the right person to assist them with an important problem. Such is often the nonlinear path of the client's buying decision journey.

If we have a better understanding of how clients buy, we can be sensitive to where the prospective client is in their thought process. And if we understand where the client is in their journey, we can do a better job of providing them with the things they need to become more confident in their decision. In doing so, we increase our chances of success at winning client business.

The seven elements framework also helps us assess how well we are doing, as individuals and as organizations. We can gauge how well we are doing at building awareness, better understand how well we are doing at helping others understand exactly what we do best, and determine how successful we are at demonstrating our professional credibility to others in ways that build respect for our capabilities.

Understanding Client Risk

Walt Shill is the managing partner and global commercial director of ERM. ERM is a leading global provider of environmental, health, and safety consulting services, employing over 5,000 consultants worldwide. Walt's as close as anyone I've ever met to being a guru in the profession of management consulting.

He has spent over 30 years in the field as a partner at McKinsey & Co., Accenture, and now in his senior leadership role at ERM. If Walt ever gets around to writing a book on lessons learned over his career, I'll be the first in line at his book signing. I met Walt a few years ago when Tom and I were conducting the research for *How Clients Buy*. One of the most insightful things I learned from Walt was the importance of understanding the risks clients face in choosing to hire us.

> One thing I always underestimated – that I now value the most – is that in many cases if you hire a consultant, it's a career risk, and the bigger the prize, the bigger the risk.

Risk is an important component in the client's buying decision journey. So much of what clients are trying to do, consciously or not, is mitigate the risks they face in choosing which expert to hire, or even if they should bring in outside help. Risks are one of the key drivers in why buying our services is so much harder than buying a product.

Here are just a few of the risks clients face:

- **Competence risk**: Are we really good at what we do?
- **Culture risk:** Are we a good cultural fit?
- **Performance risk**: Will we actually follow through on doing what we say we will?
- **Integrity risk**: Will we do what's best for the client at all times?
- **Reputational risk**: Will this hurt my firm's reputation if the project ends poorly?
- **Financial Risk**: Will this impact our firm's financial performance if things don't go well?
- **Career risk**: Will my career be derailed if this project goes badly?

Going back to our original thought experiment from the book's introduction, I think client risk is one of the key reasons why it's relatively easy to buy a home, but very difficult to choose an architect.

Buying a product, even an expensive item, feels less risky to us. Even if we can't verbalize it, we feel more secure in buying a product. We can touch it, sit in it, or walk through it. It's a real thing, and it's not hard for our brains to imagine what it will feel like in owning it.

Buying services, on the other hand, feels risky to us, even if we can't quite articulate it. This discomfort sits in the pit of our stomach. We can't easily see a person's expertise or the intentions in their heart. We often don't know. So we grasp for straws in an attempt to sort out who the true experts are, the honorable from the scammers, the reliable from the flaky.

Understanding client risk is incredibly important in helping us become more successful at winning client business. We can assist prospective clients by doing specific things that give them more confidence that:

- We are good at what we do.
- We will deliver on what we say we're gonna do.
- We have their best interests at heart.
- And, in the end, they'll have a wonderful experience in working with us.

It's Not a Long *Sales Cycle*, It's a Long *Buy Cycle*

It's often said in the world of professional services that our industry has a long sales cycle. In other words, it's not a quick sale. Some products – a new mobile phone, for example – are often a quick purchase. We walk into Verizon and an hour later we walk out with a new smart phone. It took us all of one hour to scan all of the options and find the one that best meets our needs and budget.

Professional services are not easy to buy. We need to know someone. We need to respect their expertise. We need to trust that they are honest and ethical. These are not things that can be easily discerned in one hour's time, or a day, or a week. Even if a client is highly motivated and ready to buy, they will often struggle for months in deciding which person or organization to work with.

As I've come to better understand how clients buy, and the many risks they face in choosing to work with us, I've become much more sympathetic to their situation. I understand now why it often takes a very long time to sign a work agreement: it's a *long buy cycle*. And the client is doing the best that he or she can do to pick the business partner who will maximize their chances of a positive outcome, and mitigate the risks that make them lose sleep at night.

Understanding the client's buying process helps us become better partners in their journey. And it also helps us appreciate why it's so much easier to hire us the second time. Or why a recommendation from a friend is so helpful.

Before we begin learning the rainmaker skills, we're going to first take a look at the key client pathways – where clients come from. I know, you're thinking: just tell me what I need to do to win client business. As the kung fu master says, *Patience, grasshopper*. It doesn't work if we simply jump to the "what to do" part. In order for the rainmaker skills to make sense, we have to first build a solid understanding of the consumer behavior and human psychology at play in how clients buy.

References

John Senaldi story: Interview by *How to Win Client Business* research team, 2020.

The 7 Steps of the Client's Buying Decision Journey: Tom McMakin and Doug Fletcher. *How Clients Buy: A Practical Guide to Business Development for Consulting and Professional Services*. John Wiley & Sons, 2018.

Walt Shill Story: Interview by *How Clients Buy* research team, 2017.

CHAPTER

3

Where Clients Come From

Understanding the Key Client Pathways

History doesn't repeat itself but it often rhymes.

—Mark Twain, writer

I received a knock on my office door today. It was a financial advisor from Edward Jones. Her name was Molly and she was prospecting door to door in an attempt to drum up business.

"How long have you been with Edward Jones?" I asked.

"About a month," Molly replied.

"How's it going?" I asked.

"It's going all right, I suppose. I'm just getting started," she explained.

"Have you landed any clients?" I asked.

"Yes, actually – I've landed two clients over the past four weeks," Molly beamed, "a young guy who recently graduated from college who wanted to get started saving, and a middle-aged woman who was looking to roll over her 401(k)."

"Well, congratulations," I said, "that's a good start."

Molly's story may not be that familiar to many of us; most of us don't prospect door to door. But it does fit into one of the traditional client pathways; it just may look a little different in your profession. Understanding these traditional pathways is essential to understanding the importance of the five rainmaker skills that we'll soon discuss.

The Seven Most Common Client Pathways

Over a cocktail at a dimly lit bar, I'll confess that I've spent more time than I care to admit thinking about *where clients come from* and *how we can get more of them*. Reflecting upon my own experiences and on hundreds of stories from others, I have discovered that clients arrive via seven predictable pathways:

1. Repeat business (from a satisfied client)
2. Referrals (from a satisfied client, trusted colleague, friend, or acquaintance)
3. Inquiries (from someone you know)
4. Inquiries (from someone you don't know)
5. Warm prospecting (with someone you know)
6. Warm prospecting (with an introduction)
7. Cold prospecting (with no introduction)

These client pathways are ranked by their relative success rate. The first pathway, repeat business, has the highest success rate. The last pathway, cold prospecting (as in Molly's example), has the lowest success rate. Those pathways in between are arranged in an order of descending success rate. (See Figure 3.1.)

Why does repeat business have the highest success rate? As we discussed in the previous chapter, clients hire people who they know, respect, and trust. If we've done great work for a client in the past, there is a much greater chance they'll hire us again in the future or recommend us to others.

If you're cold prospecting, the prospective client doesn't know you, and has no reason yet to respect or trust you. They may eventually choose to work with you, but you'll have to allow time for the client to move through the milestones of the client's buying decision journey.

For some of us, a prospective client may not know anything about what we do, or how our service is remotely helpful. Thus, we've got some work to do and this will take time. It doesn't mean that prospecting doesn't work in our lines of work. It can work when done right. It's just that the success rates will typically be lower and the sales cycles longer.

In a market survey my firm recently conducted with professional service providers, the data aligns with my experience and observations. As illustrated in Figure 3.2, referrals and inquiries represent roughly two-thirds of all new client business.

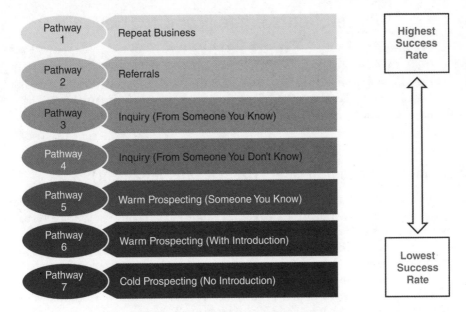

FIGURE 3.1   The Seven Client Pathways

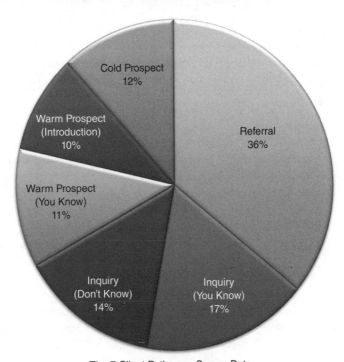

The 7 Client Pathways: Survey Data

FIGURE 3.2   The Seven Client Pathways Pie Chart

Where Clients Come From: The Seven Client Pathways

Every successful professional I speak with has stories about where their new clients come from. Each client story is different, but they have common themes. These themes are captured in these client pathways.

Pathway 1: Repeat Business (from a Satisfied Client)

This week I signed a new work agreement with a previous client for a coaching program for his consulting firm. I've known Paddy Fleming, the managing director, for about 20 years. We first met when we were both in our early 30s and my firm was doing some work with his.

At the time, Paddy was just getting started with his organization. Years later, Paddy was chosen by his firm's board of directors to be the managing director. When *How Clients Buy* was published, Paddy reached out to me to see if I would lead a client development workshop with his team. That was about 18 months ago. Last month, Paddy reached out again with a more extensive coaching program in mind.

The fastest work comes from those we have done good work for in the past. What I didn't appreciate earlier in my career is why this is so. Clients face a litany of risks they struggle to mitigate in hiring us: performance, cultural, reputational, financial, and career risks. These intangibles make it difficult in hiring us for the first time.

There are exceptions to this pathway that we'll discuss later on. Some of us provide services that a client hopes they will never need – and hopefully will never need a second time. Think about the clients of a bankruptcy attorney or a crisis PR consultant. If you do great work, hopefully the client will never need your services again. Yet, doing great work – even for once-in-a-lifetime services – can lead to referrals. This leads us to the second pathway.

Pathway 2: Referrals (from a Satisfied Client, Trusted Colleague, Friend, or Acquaintance)

What's a prospective client to do if they haven't hired someone for a specific service before? Maybe they are looking for a web designer, or a financial planner. More often than not, they will ask around to see if someone they trust has a recommendation.

Referrals are the leading source of new business according to my survey, representing 36% of all new clients. This is two times higher than any of the other new client pathways. If we come highly recommended by someone who knows us, we've moved from home plate to second base, maybe even third. In the absence of a repeat client, a strong referral is the next fastest pathway.

The referral source is split roughly 60/40 between our client base (57%) and our professional network (39%). The leading source of referrals is from those we are currently working with or have worked with in the past. This, again, highlights the importance of doing great work.

The next leading source of referrals is from our professional network, or ecosystem: professionals outside of our firm with whom we have a close relationship. This ties into the importance of Rainmaker Skills 3 and 4: building your professional ecosystem and developing trust-based relationships. Chuck McDonald – an attorney in Columbia, South Carolina, specializing in serving the needs of clients in the construction industry – puts it this way:

> The best way to sell your services is not for you to sell your services, but to have someone else sell your services. Nobody wants to hear you say how good you are. It means so much more to have someone they trust say how good you are. Unless you screw up a good referral, you're going to get the business.

Pathway 3: Inquiries (from Someone You Know)

The next most common pathway is an inquiry from someone we know but have never worked with previously, similar to my friend John Senaldi's story from the previous chapter. Our survey data suggests that this pathway is the second leading source of nonrepeat client business at 17%, roughly half that of referrals.

The client's buying decision journey is often nonlinear. We may meet someone, get to know them, they come to respect and trust us, but they are not in the market for what we do. Years later an opportunity arises where they find themselves in need of outside help, and you are top of mind. In this scenario, the risks a prospective client faces in hiring us are mitigated by the strength of our existing relationship.

Sarah is a friend of mine who is a young partner with one of the big four accounting firms. She loves classical music and serves as the treasurer on the nonprofit board of her city's symphony. Richard, one of her fellow symphony board members, is the CFO of a Fortune 500 consumer products company. Sarah and Richard have known each other for about three years since joining the nonprofit board, and work closely together on the board's audit committee.

About a year ago Richard began searching for a new accounting firm to handle his company's global audit responsibilities. He narrowed his search down to three firms and Sarah's was on his short list of candidates. After a rigorous evaluation process, Richard and his team selected Sarah as their new audit partner.

When we have an existing relationship with someone, and that relationship is characterized by mutual respect and trust, we put ourselves in a stronger position to win the client's business. When we get a call from someone with whom we have an existing relationship, the client's buying decision journey has fewer obstacles. In the absence of these obstacles, our success rate will be higher.

Pathway 4: Inquiries (from Someone You Don't Know)

It's wonderful when we get a call from someone interested in what we do. It's a pleasant surprise when it's from someone we don't know. Our survey data suggests that this pathway represents about 14% of our new clients.

It can be very hard for prospective clients to tell when we're very good at what we do. One of the reasons for this is information asymmetry. Information asymmetry exists when there is a gap – sometimes a large gap – between what you know about your work and what a client knows. Because clients are not often experts at the type of help they need, it's usually difficult to identify the truly talented. Many in our professions may look the part, talk the part, and act the part.

In some professions, there are credentials that serve to indicate that a person is capable of doing the work. Law, accounting, architecture, engineering, and medicine are a few examples. If we're looking to hire an accountant, we wouldn't hire someone without a CPA qualification. Similarly, we wouldn't hire an attorney or doctor who wasn't licensed to practice their work.

In other professions, it's not so easy. Management consulting, IT services, and marketing/advertising, don't have the same standards. Anyone can claim to be a marketing expert or web designer or strategy guru. What clients need are credibility markers. "Credibility markers" is the term that academics have given to things that demonstrate our capability.

Often a prospective client discovers us through things we've done in the past. It could be a recent article you wrote in a trade journal, or a speech you gave at a conference. Perhaps you were highlighted in the news about a recent project, such as a prominent new building on the city's skyline. Or maybe someone discovered you via your website because of a white paper or blog post.

Client work sometimes comes from people reaching out to us as of result of these credibility markers. In the absence of a repeat client, referral, or an inquiry from someone we know, this client pathway is the next leading source of business. The importance of credibility markers leads us to a better understanding of the second rainmaker skill, demonstrating your professional expertise.

Pathway 5: Warm Prospecting (with Someone You Know)

The last three pathways are with people or organizations who haven't sought our help. Therefore, we don't know their level of interest, ability to hire us, or readiness relative to their other priorities.

Pathway 5 is when you explore business opportunities with someone where there is an existing relationship. Maybe it's someone you recently met at a conference, or someone you've known for many years. Our survey data suggests that Pathway 5 represents 11% of new client business – significantly less than referrals and inquiries, but for many professionals an important pathway.

Prospecting is always going to be harder than repeat business, referrals, or an inquiry from someone actively seeking help. There are a host of headwinds we face when we are the one proposing the idea. These headwinds include:

- *Caveat emptor:* Buyer beware; our defenses are up when someone wants to sell us something.
- *Status quo:* Human nature being what it is, staying the current course is often perceived as the safer route.
- *Buy-in:* Others in the organization aren't on board with us or the idea.
- *Budget:* No funding is available at this time.
- *Timing:* Too many conflicting priorities exist.

Pitching an idea to someone we have a relationship with is going to be easier than pitching an idea to a stranger. When an relationship exists based upon mutual respect and trust, we have already achieved three of the milestones in the client's buying decision journey: awareness, respect, and trust. This is why building your professional ecosystem and developing trust-based relationships are two of the key rainmaker skills.

Prospecting can lead to new client business. It takes time, and you'll have plenty of swings and misses. Prospecting is not for everyone. But for those who learn the skill, it can be a highly effective pathway to new business.

Pathway 6: Warm Prospecting (with an Introduction from a Mutual Friend)

Pathway 6 is when you're prospecting with someone to whom you've been introduced, and represents 10% of new clients according to our survey data. The prospective client hasn't asked for help, and she may not be actively in the market for your service. But based upon your research, you feel you can help her.

You have a friend or colleague who knows the prospective client, and you've been kindly offered an introduction. Warm introductions are a form of currency shared among friends and colleagues in the business world. And they grease the skids of commerce in a karmic, pay-it-forward kind of way.

When appropriately used, warm introductions help break down barriers of human resistance. When we're introduced by a mutual friend, we're much more likely to get to spend some time with a prospective client. For many, accepting a warm introduction is a social courtesy that we extend to a mutual friend.

Mark lives in San Diego and is a senior director for a company that provides services to law, accounting, and financial firms. Mark and I worked together about 20 years ago and have kept in touch over the years as we progressed through our careers.

Mark and his team have been researching Bank of America for several months, and feel they are a good fit. His firm does similar work for several large banks in New York. Over beers one warm evening in La Hoya, Mark asks me, "Do you know anyone at Bank of America?"

I tell him, "Yes, actually, one of my college friends is a senior vice president in their Atlanta office." And, because I respect and trust Mark, I am happy to provide an introduction.

Six weeks later, Mark and my college friend at Bank of America have dinner together at the Buckhead Diner in Atlanta. Maybe they will end up doing business together, or maybe not. Time will tell. But it would have been much harder for Mark if it weren't for the warm introduction I provided. An introduction from a mutual friend serves as a social bridge between two strangers. This bridge helps to break down barriers that often exist between two people who have never met.

Pathway 7: Cold Prospecting (with No Introduction from a Mutual Friend)

Cold prospecting refers to efforts on your part at meeting someone with whom you would like to do business – without the benefit of an introduction. Cold prospecting represents roughly 12% new client business, according to our survey data.

Many successful rainmakers have built their careers off of cold prospecting. Cold prospecting can work, it's just that it takes longer to build a relationship with someone who doesn't know you and isn't introduced to you.

There are two types of cold prospecting. The first is blind prospecting. Molly from Edward Jones was blind prospecting, knocking on doors without any prior research on the prospect. The better and more effective form of prospecting is what is called targeted prospecting. Targeted prospecting is when you have a very specific target audience and have identified a good candidate based upon extensive market research, as was the case with my friend Mark in the previous story.

When you've done your homework, and you have identified an organization or an individual who you would like to serve, your best option whenever possible is a warm introduction from a mutual friend. But sometimes we simply don't know anyone who can introduce us. In these cases, cold prospecting may be our only option.

Ask anyone who cold prospects for a living and they'll readily admit that it's hard work, and you'll strike out much more often than you'll score. You have to develop thick skin and get used to a lot of rejection. The key is to find common ground with your targeted prospect, something you have in common that serves as a proxy for an introduction from a mutual friend.

Connecting the Dots

These seven client pathways represent the most common ways in which we land our client business. Each client journey may be as unique as a snowflake, but they often share many similarities.

Hopefully you're beginning to connect the dots between the client's buying decision journey and the five rainmaker skills. Laying this foundation of consumer behavior and human psychology helps to better understand why the rainmaker skills are so effective.

Now that we're armed with an understanding of *how clients buy* and *where clients come from*, it's time to roll up our sleeves and learn the five rainmaker skills. Before we do, there's one last topic we need to discuss. This is the last thing we need to cover before learning the rainmaker skills – I promise.

What if we don't want to sell our services? What if we self-identify as an introvert? Aren't all rainmakers highly sociable extroverts? The short answer is no. My experiences and those of many other highly skilled rainmakers indicate that there are many ways to win client business; unique approaches tailored to our own strengths and preferences. Let's squash this myth of the extroverted rainmaker before we go any further. If we're to become successful at winning client business, we have to first understand that there is no *one right way*.

References

Molly from Edward Jones story: Inspired by real characters known by the author.

The Seven Most Common Client Pathways: *Survey conducted by Fletcher & Company*, LLC, 2020. https://www.fletcherandcompany.net/new-client-pathways-survey-findings/.

Paddy Fleming story: Interview by How to Win Client Business research team, 2020.

Chuck McDonald story: Interview by How to Win Client Business research team, 2020.

Sarah and Richard story: Inspired by real characters known by the author. Names and locations have been changed.

Mark story: Inspired by real characters known by the author. Names and locations have been changed.

CHAPTER

4

Rainmaking for Introverts and People Who Don't Want to Sell

Winning Client Business While Being True to Yourself

At its core, the idea of purpose is that what we do matters to people other than ourselves.

—Angela Duckworth, author, *Grit: The Power of Passion and Perseverance*

Carl Jung, the father of modern personality theory, first introduced the world to the concept of introversion and extroversion in his 1921 classic *Psychological Types*. According to Jung:

Introverts are drawn to the inner world of thought and feeling, extroverts to the external life of people and activities.

If we accept the premise that we have to become rainmakers if we are to become partner – or for our solo practice to thrive – this creates an interesting conundrum for a big swath of us. Marketing ourselves is viewed as the realm of the outgoing, the charismatic, those with the gift of gab. There's a reason why we think this way. And to understand why, we need to go back over 100 years.

According to cultural historian Warren Susman, Jung's theories spawned a shift in early twentieth-century American societal values away from a Culture of Character to a Culture of Personality. In the Culture of Character of the 1800s, the ideal self was serious, disciplined, and honorable. In the emerging Culture of Personality of the twentieth century, Americans became captivated by those who were bold and entertaining.

This shift in American thinking is exemplified by the success of Dale Carnegie. Born into humble beginnings in rural Missouri in 1902, Dale went on to become a best-selling author and leader of the emerging American belief that the world is ruled by extroverts. According to Susman:

> Carnegie's metamorphosis from farmboy to salesman to public-speaking icon is also the story of the rise of the Extroverted Ideal. Carnegie's journey reflected a cultural evolution that reached a tipping point around the turn of the twentieth century, changing forever who we are and whom we admire, how we act at job interviews and what we look for in an employee, how we court our mates and raise our children.

America's societal views on extroversion and introversion are evolving; the pendulum is swinging to a more balanced perspective. Susan Cain, a leading twenty-first-century researcher on introversion, offers this perspective in her wonderful book *Quiet: The Power of Introverts in a World That Can't Stop Talking*:

> We make a grave mistake to embrace the Extrovert Ideal so unthinkingly. Some of our greatest ideas, art and inventions – from the theory of evolution to van Gogh's sunflowers to the personal computer – came from quiet cerebral people who knew how to tune in to their inner worlds and the treasures to be found there.

Furthermore, according to Cain, one-third to one-half of Americans lean toward introversion – in other words, one out of every two or three people you know. If these statistics surprise you, that's probably because so many people pretend to be extroverts. Cain's research suggests that closet introverts pass undetected in the corridors of corporate America.

I suspect that a healthy percentage of us in the professional services tilt toward introversion. By nature, many professionals are cerebral, inward-focused problem-solvers. We live inside of our heads much of the time. We draw energy by wrestling with thorny issues for hours, days, or weeks on end. Many of you may relate to what I'm talking about.

The Rainmaker Mindset

You may be thinking to yourself, "Wait, I don't want to be a salesperson!" If so, I'm confident you're in good company. Don't panic; we don't have to be salespeople at all. Or fake extroverts. The key is to view our work as rainmakers as helping others; to see the process of client development as a journey in helping identify and find solutions to client problems.

Mindset Shift One: Seeing ourselves as problem-solvers instead of salespeople

Some may see the distinction between calling us *problem-solvers* versus *salespeople* as purely semantics. It is far more than this. It is a total mindset shift in the way we view the work we do as rainmakers. It's important to see our role as partners in a journey of helping solve important problems rather than selling services. Products may be sold by salespeople, but clients hire trusted advisors.

Tim Nath is the operations practice director of Aspirant. Aspirant is a Pittsburgh-based management consulting firm with approximately 100 employees. Like many of the senior professionals at Aspirant, Tim comes from an operating background. Tim shared with me that when he first entered consulting he was uncomfortable with selling:

> If I could go back in time, I would like to know that rather than going into a client with "Here's what I can sell," it is really about "What problems do you have that I can help solve?" I'm an introvert by nature. It always rubbed me the wrong way to think about being in a "sales position." But when I think about it as "I'm helping someone solve an important problem" – that's something that I can get on board with.

There's an old adage in sales that goes: *no one likes to be sold, but everyone loves to buy.* This is to say that clients don't like pushy salespeople, but they welcome genuine help from trusted advisors who can help them solve problems and advance their goals.

Wherever you land on the introversion/extroversion spectrum, I encourage you to follow the wisdom offered by the most successful rainmakers: *whatever you do, don't call it selling*.

Lessons from Dominic Barton, Global Managing Director at McKinsey & Company

Dominic Barton successfully led McKinsey & Company from 2009 to 2018. While McKinsey is one of the most well-known and respected management consulting firms in the world, Dominic is not a household name like Elon Musk. We won't see Dominic on *Shark Tank* like Sara Blakely, but in the world of management consulting, Dominic casts a large shadow.

Dominic shares with us his thoughts on winning client business:

> Relationships with clients are critical. These relationships are based on trust... the way you've worked together...how you interacted together with people – that becomes the most critical thing.

What are the implications of this if we're more inward-looking than outgoing? Clearly, we need to know people and build our network of professional relationships – and this is where extroverts excel, right?

There is hope for us introverts, because as we'll learn from Dominic, there is no one right way to build our personal networks. With over 25,000 employees, I'm guessing McKinsey has more than a few introverts in its mix. Dominic shared this:

> It's very important to build a professional network. I didn't realize this until I was five or six years in the firm. What is the approach you're gonna use to develop relationships with different people? There are many different ways of doing it.

Dominic continued:

> That can be through knowledge. One of my mentors established a banking practice. All his client relationships came from what he wrote. He was not a guy who did cold-calling. It was through his writing. He had perspectives on the industry, where it was going, what he thought needed to be done and how. And so his model was he wrote a lot. And through writing, he established a reputation and then a network. And he was always busy.

This story is full of hope for many of us. Many writers lean toward the introverted side. We spend a lot of time reading, thinking, and writing about our perspectives on matters pertaining to our areas of interest.

Writing, as Dominic's story suggests, is a form of networking – networking for introverts, a way of becoming friends with strangers. The author Pico Iyer once quipped, "Writing is, in the end, that oddest of anomalies: an intimate letter to a stranger."

Pico's sentiment is shared by Judy Selby, a partner in the New York law firm Hinshaw & Culbertson LLP. Judy has over 25 years' experience in complex insurance coverage litigation and international arbitration. Judy writes a lot for her target audience on the topics of cyber insurance and privacy, and she has found writing to be an effective way for her to build relationships with those she wishes to serve.

Judy offered this from her writing experience: "When I meet people for the first time, they feel like they already know me. They're like, 'Judy Selby – yes,

I know you – I've read your work. And, so, you come into this new relationship with credibility and trust."

Another example from Dominic Barton offers hope to those of us who don't shine at business happy hours:

> And there was another mentor of mine. He also wrote, but his approach was more personal. He used cold-letter writing. He would think about transitions for new CEOs coming in – and these were people that he didn't know. And he would write the CEO and say, "I'd like to talk to you because I've got some news on what you might be thinking about, and I'd like to have a discussion."

Susan Cain, the author of *Quiet*, makes a point of differentiating between introversion and shyness. According to Cain, "Introverts are not necessarily shy. Shyness is the fear of social disapproval or humiliation, while introversion is the preference for environments that are not overstimulating." I think this is an important observation. Some introverts are shy, but many are not. An example Cain offers is Bill Gates: "Bill Gates, who by all accounts keeps to himself, is unfazed by the opinions of others."

One last example from Dominic offers an approach that I think many of us will feel more comfortable with in building our personal network:

> It was probably when I was four or five years in Toronto and the office manager asked me to help with a hospital board. And I said, "Why are you doing this to me? I'm not even interested in healthcare." And the office manager said, 'Look, it's gonna be good for you. You've gotta learn. If you wanna be here for the long-term, you've gotta understand how boards work and what CEOs have to deal with.
>
> I actually learned a lot about how boards work. And I got to know the board chairman really well, who turned out to be a leading business guy. And the chairman says to me one day, "Well, you know, you should come and talk to me about my business." So, the office manager was exactly right. It was good for me. I learned how boards work. And I built a relationship with the chairman from the nonprofit board who eventually became a client of mine.

Dominic's stories highlight the fact that there are many successful ways we can meet people whom we wish to serve. I have met many successful rainmakers who were introverts. I have also known many who were extroverted as well.

The key to becoming a successful rainmaker is being true to our unique personality and choosing approaches that play to our strengths and preferences. Whether introverted or extroverted, our success as rainmakers begins with a desire to help others.

References

Introversion/extroversion: Susan Cain. Quiet: *The Power of Introverts in a World That Can't Stop Talking*. Crown Publishers, 2012

"Prof. Warren Susman, Rutgers Historian, Dies." *New York Times*, April 22, 1984.

Tim Nath story: Interview by *How to Win Client Business* research team, 2020.

Dominic Barton story: Interview by *How Clients Buy* research team, 2017.

Judy Selby quote: Interview by *How to Win Client Business* research team, 2020.

The Five Skills We Must Learn If We Want to Become a Rainmaker

Skill 1: Create Your Personal Brand Identity

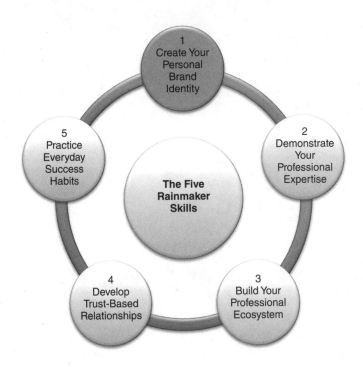

CHAPTER

5

Decide What You Want to Be Known for and Who You Wish to Serve

You Can Be Known for Anything, But You Can't Be Known for Everything

Regardless of age, regardless of position, regardless of the business we happen to be in, all of us need to understand the importance of branding. We are CEOs of our own companies: Me Inc. To be in business today, our most important job is to be head marketer for the brand called You.

—Tom Peters, author, *The Brand Called You*

We have laid the groundwork of understanding the client's buying decision journey: what goes on in the minds of clients, how they think, and the risks they face in choosing us. We discussed the most common client pathways: how clients come to discover us, respect our abilities, and trust that we have their best interests at heart.

We also learned there is no one rainmaker personality or style – and we can adopt an approach that best fits our strengths and preferences. With this foundational knowledge in place, we are now ready to begin exploring the five skills required to become a rainmaker:

- Skill 1: Create Your Personal Brand Identity
- Skill 2: Demonstrate Your Professional Expertise
- Skill 3: Build Your Professional Ecosystem
- Skill 4: Develop Trust-Based Relationships
- Skill 5: Practice Everyday Success Habits

Creating your personal brand identity is the first rainmaker skill. You may be scratching your head. "Wait, what? Branding is for big companies like Google, Apple, and Nike, right?" Sure, big companies spend an incredible amount of effort building their brands. Corporations know that branding helps establish relationships with their customers. And the stronger the relationship, the more likely consumers will choose their products.

The concepts of branding are also especially important to us as individuals. We may be experts at finance, organization dynamics, or public relations. But regardless of our profession, we are all in the relationship business. And having a strong brand identity is the first step in connecting with those we wish to serve.

Your brand image is the perception held in someone else's mind about who you are, what you do, and how you do it. Each of us already has a brand identity, whether it's been thoughtfully crafted or happened by chance. The key difference between rainmakers and the rest of us is that rainmakers carefully curate their brand identity. It's not left to chance. Establishing your personal brand identity is choosing what you want others to think of first when they hear your name.

Communicating our personal brand is not about creating a false narrative of who we are. The essence of great branding is about becoming more of who we are and who we aspire to be. It's communicating to our target audience who we truly are. It's a process of defining the most authentic versions of ourselves in a way that best positions us for success.

Before we can begin to present ourselves to those we wish to serve, we have to first establish what it is we want to be known for. In these next four chapters, we'll discuss what this means, why it's important, and how to do it successfully.

The Birth of Personal Branding

The term "personal branding" was first coined in the late 1990s by management guru Tom Peters, during a period when employees were recovering from the downsizing of corporate America. Because large employers could no longer be counted on as a secure source of employment, individuals began to realize the importance of taking ownership of their own career. Magazines such as *Fortune*, *Wired*, and *Fast Company* shifted the focus of attention from big corporations to the individual.

Michael Port, the *New York Times* best-selling author on personal brand building, offers this perspective:

> Brands are not just for big corporations. In fact, a personal brand will serve as an important key to your success. A personal brand will help clearly and consistently define, express, and communicate who you are.

Many of us think of branding as cleverly designed logos and catchy marketing. But this is something different; we're talking about your identity and reputation – what you are known for being good at wrapped in the coat of your personality and character.

Port adds, "Personal branding is far more than just what you do or what your website and business cards look like. It is you – uniquely you. It allows you to distinguish yourself from everyone else: what is unique about who you are and what you do."

Your personal brand identity is a complex blend of many factors that represent you to the world:

- **What you do** (your expertise)
- **Who you know** (your personal ecosystem)
- **Your experiences** (what you've done)
- **Who you wish to serve** (your target audience)
- **How you can help** (the problems you solve)
- **Your personality** (funny, serious, outgoing, shy, etc.)
- **Your character** (trustworthy, generous, caring, kind, etc.)

To be able to assist a client, they have to know us and understand how we can help them. This directly connects with milestones one and two of the client's buying decision journey: awareness and understanding. The key to being remembered is establishing your personal brand identity.

In the upcoming chapters, we'll focus on the more tangible aspects of your brand identity: what you do, who you wish to serve, and how you can help. Later on, as we explore the other rainmaker skills, we'll examine the more intangible aspects of your brand: your personality and character.

Stupid Choices!

When our daughter, Abby, was entering high school, I gave her this fatherly pep talk one night before bedtime: "Get involved in school activities," I suggested. "It's important to be a part of things, and not just stand on the sidelines. You'll make good friends, and have more fun in high school."

I continued, "It doesn't matter what you choose to get involved in. It can be sports, music, theater, debate, art, civic clubs...you name it. Find something that interests you and go do it."

Thinking back, that was pretty good advice. As all parents learn, our kids often ignore our well-intentioned words of wisdom. On this occasion, however, my daughter took what I offered to heart.

By the second week of high school, Abby wanted to sign up for everything. She wanted to play sports, join the band, participate in the school play, learn photography, and join a civic club. Over dinner I said to her, "I can tell you really took my advice seriously, and I admire your enthusiasm. But you can't do all of these activities, at least not all at once. There's not enough time to do all of these things. You can pick one activity per semester," I continued. "Between school and all the other things you've got going on, there isn't enough time to do them all."

She sighed, "All right then, I'm most interested in playing sports." Abby went on to play hockey and lacrosse. She thrived in a team sports environment; she made many great friendships and learned many life lessons. Our daughter's experience led to what became known in our family as Dad's mantra: *You can do anything, but you can't do everything*.

This reminds me of another parenting story, this one involving our son, Duncan. Each night after dinner, we would allow our kids to have a small dessert once they finished their meal.

I can vividly remember this one night with Duncan when he was about two. We said, "OK, you get to pick your treat. What'll it be: a cookie or a scoop of ice cream?" Our toddler's face began to turn red. He scrunched up his tiny nose, pondering the choice he had to make. Finally, after about a minute of deep thought, he had a bit of a meltdown. Through tears, he proclaimed, "Stupid choices!"

We all have moments that become family legends. These stories are two of ours, and they have made us laugh many times over the years. There is a message in these stories as it relates to establishing our personal brand identity: in our career we have to make choices about what we want to be really good at and known for.

Having to choose one thing to be known for can create discomfort for some, and you are welcome to shout at the top of your lungs, *Stupid choices*! But, with a twist on our family mantra: *You can be known for anything, but you can't be known for everything*.

There are choices to make in life – sometimes hard choices. In order to gain something of value, we have to be willing to give up something. It's a hard fact of life. The same is true in our career as it relates to our personal brand identity.

Society's Hall Pass

I began teaching at our local university when I sold our consulting firm and had some free time on my hands. An acquaintance of mine was the associate dean at our local university's college of business. Over lunch one sunny afternoon in a café a block from campus, he suggested that I teach a class at the university.

"I have an opening right now for a class this fall semester," he said. "I think you'd really enjoy it."

"I don't know," I hesitated.

"Come on," he insisted. "How bad can it be? And, besides, what else are you gonna do with your spare time?"

"Well, I'm not lacking for things to do," I assured him. "But I'll give it some thought."

I have been teaching now for over six years. Choosing to teach was one of the best decisions I have made in my career. I was correct in my assumption that it takes a lot of work. However, I underappreciated how much I would enjoy sharing my passion for business with students, and helping them navigate the beginning of their careers.

The course that I teach, the Capstone Course on Business Strategy, is comprised of seniors in their last semester before graduation. It's a great time to be a part of their lives, when they're excited to be graduating from school and anxious about transitioning to a career.

During the semester I offer a few pieces of career advice. I share with them the idea that anytime someone offers you advice, replace the concept of *advice* with the word *opinion*. "That's all advice is, really." I tell them. "Someone's opinion."

During my career advice talk, I tell them about my concept of Society's Hall Pass. In my experience, society gives young adults in their 20s a hall pass on settling into a specific career path. Society doesn't share this information, but parents and employers expect that young professionals will explore and experiment with their careers. We, too, were once their age and can relate to their uncertainty.

The advice I offer my seniors is this: take advantage of Society's Hall Pass. Try different jobs, experiment with new industries, and test out different companies. I clarify that I'm not suggesting that they change their minds every month.

But I propose that if you're in the same job doing the same work for two years straight, you need to find a new opportunity. By the time you reach your 30s, you should have had three or four totally different experiences.

Why do I offer this advice to young adults? Because Society's Hall Pass expires once you reach your mid-30s. At some point parents, bosses, and work colleagues expect that you'll find something that you are willing to stick with. If you continue to hop jobs, careers, companies, and industries well into your 40s, society begins to label you as a flake. It may be unfair and cruel, but, in my experience, that's just the way it is.

By the time you reach your mid-30s, society expects that you'll find your niche and begin to develop a degree of skill and competence at something. Perhaps not surprisingly, it's at about this time that we begin to think about the partner track or starting our own firm.

I offer you this ~~advice~~ opinion: if you are in your 20s and in the early phase of your career, skip the hard decisions for now on establishing your personal brand identity. Instead, work hard, do the very best you can at the work you have, and try to learn as many things as possible. There will be plenty of time for you in 5–10 years to get down to the hard decisions of choosing your niche.

If you're in your 30s and aspire to become a partner – or have hopes of one day owning your own firm – it's time for us to dive deeper into why you need to establish your personal brand identity.

A Short History of Branding

What we know of today as branding began over 100 years ago with a pamphlet written by James Walter Thompson, the founder of the now famous New York ad agency bearing his name. In 1900, Thompson began encouraging manufacturers to advertise directly to consumers with strongly branded messages. This was the early commercial beginnings of what we now recognize as modern branding.

But branding began long before the twentieth century. Historians believe that the concept of branding began in ancient Egypt about 2700 BCE. A brand communicated information about a product's origin, ownership, and quality.

Over time, the practice of branding was adopted by artisans, farmers, and traders throughout Africa, Asia, and Europe. Seals, which were a form of brand, are found on Chinese products of the Qin Dynasty from around 200 BCE. An example of one early Chinese brand is that of *White Rabbit* sewing needles of Jinan Liu's Fine Needles Shop.

A maker's mark was later used by potters in ancient Greece and Rome to identify the artist. These marks offered information about place of origin, type of product, quality claims, and the name of the manufacturer.

In the European Middle Ages, maker's marks were added to a wide range of goods. Some brands still in existence today began in the 1600s. One of the most famous is the distinctive Bass & Company beer logo, which the company applied to casks of its famous pale ale. In 1876, its red-triangle logo became the world's first registered trademark issued by the British government.

References

Personal branding: Daniel J. Lair, Katie Sullivan, and George Cheney. "Marketization and the Recasting of the Professional Self," *Management Communication Quarterly*, 2005.

Michael Port. "How to Develop a Personal Brand Identity." *Huffington Post*, 2011.

Dave McNally and Karl D. Speak. *Be Your Own Brand: A Breakthrough Formula for Standing Out from the Crowd*. Berrett-Koehler Publishers, 2011.

Tom Peters. "*The Brand Called You*." *Fast Company*, 1997.

Abby and Duncan story: Inspired by real characters known by the author.

Society's Hall Pass: From author's lecture notes, Senior Capstone Course on Business Strategy, BGEN499, Montana State University, 2017.

A Short History of Branding: Sladjana Starcevic. "The Origin and Historical Development of Branding and Advertising in the Old Civilizations of Africa, Asia and Europe." *Marketing* 46 (January 2015): 29–46.

CHAPTER

6

The Power of Focus

The Key to Being Remembered

> The narrower and more focused your brand, the easier it is for people to remember who you are.
> —Goldie Chan, writer, *10 Golden Rules Of Personal Branding, 2018, Forbes*

The first step in the process of becoming a rainmaker is establishing your personal brand identity; it's the cornerstone of everything else you'll do to win client business. The primary reason a highly focused personal brand works so well is that it helps you stand out from the crowd. Until those we wish to serve remember us, there is very little chance that our talents will have a chance of being discovered and utilized.

Of the top client pathways, the top two – referrals and inquiries from people you know – represent over 50% of all new client business. People need to remember you in order for these two pathways to open and flow. Having a strong focus is the key to being remembered.

Focusing your personal brand is a hard topic for many to wrap their mind around. The difficulty stems from two main reasons:

- Reason 1: The very notion of narrowing our focus defies what our intuition is telling us.
- Reason 2: It is often really hard to choose one thing to focus on.

Trevor is a highly skilled and talented project leader working for a management consulting firm in Dallas. Trevor has one client that he works with full-time – a very large technology company. Trevor's client loves him. He and his team do great work. He is seen as the person you go to when you have a difficult situation. Trevor has a knack for getting things done. Over time, Trevor built quite a brand reputation with his one client, and receives a lot of referrals within the client's organization.

Trevor aspired to become a rainmaker at his firm, and he reached out to me to see if I could assist him.

"How can I help you?" I asked Trevor.

"All of my work is with one organization," Trevor replied. "They are a great client. But I have never sold anything before. The work just keeps coming to me from my one client. I'd like to learn how to sell my work to other organizations. I want to branch out and land a few other clients. The problem is I don't know where to start."

I continued, "What types of problem do you want to help other client organizations solve?"

Trevor paused, "That's the problem. I don't have a specialty. I am a generalist. My one client comes to me with complex problems and we roll up our sleeves and help solve them. But I don't really think I'm an expert at anything. I can help solve just about any problem."

"Who do you wish to serve…outside of your one client?" I asked.

Trevor responded hesitantly, "We can help any organization, really. They just need to be rather large…say, Fortune 100. If they're not large enough, they won't have the budget to hire us."

Trevor's story is one I've heard many times. In my experience, most of us – including many high-achievers such as Trevor – have a very difficult time answering the following questions:

- **What do I do best?**
- **What do I want to be known for?**
- **Who do I wish to serve?**
- **What makes me unique?**

Many talented people have been so busy for so many years that they haven't taken the time to reflect upon what they are really good at – or what they wish to be known for. They are just really good at solving the problems that are placed right in front of them.

The problem arises when you want to take your skills to a broader audience. These new organizations or individuals don't know you. They don't know your reputation. They don't know what your expertise is. They don't know if you are dependable. They don't know if you're honest. And choosing to hire you for the first time is often a long, arduous process filled with many risks – real and imagined.

Before we begin to create our personal brand identity, it's important to understand why this is so important. Understanding why we must establish a unique identity will give you the courage to actually do it. I say *courage* because the process is filled with uncertainty and anxiety – and many professionals avoid making the hard decisions about what they want their personal brand identity to be.

If You're Not in the Client's Top Three, You're Not Going to Be Hired

There's a business concept that is crucial to understanding the importance of having a focused personal brand identity: it's called TOMA. TOMA stands for *top of mind awareness*. TOMA means the first brands that come to mind when a customer is asked about a specific product or service category. I first learned of TOMA nearly three decades ago in Dr. Paul Farris's first-year marketing class at Darden Business School. Dr. Farris explained that most customers will choose one of three brands that first pop into their minds when considering a purchase.

I recently spoke with Eric Gregg, founder and CEO of ClearlyRated – the Portland, Oregon-based industry leader in measuring client satisfaction for professional services. Eric is a prominent thought leader on the topic of measuring client service and buyer perceptions.

Eric shared this key finding from his company's research:

> Buyers usually start out aware of 4–5 firms, meet with 3, and narrow it down to one of two options.

ClearlyRated's research data supported what I had learned from Dr. Farris nearly 30 years ago: if you're not in the top three being considered for a client's business, the odds of winning are very slim.

Two of the first individuals to identify the importance of TOMA were the legendary ad men Al Ries and Jack Trout. They observed in the 1970s that one of the biggest problems brands face is standing out in an overcommunicated society: there is so much advertising coming at us that it all becomes background noise and we tune it out. In their seminal work on brand advertising, *Positioning*, Ries and Trout explain:

> The typical consumer is overwhelmed with unwanted advertising, and has a natural tendency to discard all information that does not immediately find an empty slot in the mind.

And that was even before the 24/7 news cycle, internet, smart phones, or social media. Fast forward 50 years, and you can imagine how hard it can be for a brand to break through the daily onslaught of media noise.

Ries and Trout discovered one of the secrets of achieving TOMA: *positioning*. Their main message centers on the psychology of standing out in an overcommunicated society. In order to be remembered, we have to find an empty slot in our customer's mind and fill it.

If there's already a brand parked in that spot, you need to find a new slot because otherwise you're not easily getting in. It was Ries and Trout who coined the oft-repeated but hard to follow advice: shrink the pond until you're a big fish, and then grow your pond.

Ries and Trout offer this advice:

> Narrow the focus. This is the most difficult step of all, because it is counterintuitive. Most managers and entrepreneurs look for ways to expand their product offerings. What is difficult is selecting the one concept to focus on. Most companies don't want to sacrifice. They'd rather have a handful of horses in the race to the future. It sounds right, but it doesn't work.

The idea of positioning has influenced the thinking of many thought leaders over the subsequent decades. One example is Michael Porter, the famous strategy professor at Harvard Business School. Every business student since the 1990s has learned Porter's Five Forces model for analyzing the strength of competition in an industry.

I've followed Porter's work since my days at graduate school. I've read every one of his books – and they are not what you might call light reading. If you look carefully at Porter's theories, you will see the fingerprints of Ries and Trout throughout.

A more recent example of Ries and Trout's influence is in the work of Blair Enns. Enns wrote an insightful book titled *The Win Without Pitching Manifesto*. Enns echoes the power of focus throughout his manifesto. Here's one example:

> We must recognize that as individuals we are inclined against the narrow focus that drives deep expertise, but we must also recognize that our business must have this focus if it is to prosper.

Too often, we decide to not decide and so, in our minds, leave open the possibility that we may continue to do all things for all types of clients. The avoidance of focus remains the root cause of most business development problems.

Establishing a focused personal brand identity is the key to owning a slot in the client's mind. Without it, we won't be remembered. And if we're not remembered, then we won't be on the short list when a client is ready to buy. Nor will we be remembered when a colleague is asked for a recommendation.

Filling an Empty Slot in the Client's Mind

Filling an empty slot in the client's mind is a daunting challenge for most of us. What slots are empty? What slot do I want to fill? How will I choose the right slot? The good news is – as researchers have discovered – the human mind has room for three brands for each slot.

You have to carefully pick a slot where you can be top three if you want to have a strong chance of being chosen. If you're not one of the top three brands in that category's slot, then you're not going to be remembered – and as a result, you're not likely to be hired.

Pick a category for any product or service – let's say European luxury sedans. If you're not Mercedes, BMW, or Audi, the odds of getting the sale are not good. Thinking of having an American light beer after mowing the lawn on a hot August afternoon? Hmmm, will it be a Bud Lite, Coors Lite, or Miller Lite?

OK, this same logic applies to each of us in our lines of work. Sure, we're not making cars or beer, but the human brain being what it is, it has room for only three brands, possibly four, for whatever it is you do. If you're not at the top of the list, you're not going to have a seat at the table.

If you're number five or ten in the client's mind, you're not going to have a chance to show how talented you are. That's just a fact of consumer behavior. It's why top-of-mind awareness is so important to your personal brand. So, what do we do if we're not in the client's top three? We have to narrow our focus.

The hardest part for many in establishing a personal brand identity is choosing a focus area, a specialty, or a go-to expertise. Part of this problem is that it seems to defy logic. If we narrow our focus, naturally we narrow our target audience. And if we narrow our target audience, obviously we narrow our business opportunities. Why would we want to do that?

This logic of *we can do anything* and *we can serve anyone* seems right to us. The problem is – it's wrong. In fact, it's so wrong that many of us never choose a focus area. This is partially to explain why there are relatively few among us who are highly successful at becoming rainmakers.

Reinventing Your Personal Brand

Some of us find our thing and stick with it forever. I admire these people – especially the ones who are genuinely happy. But many of us – maybe most – will reinvent our personal brands a number of times during our careers.

Choosing our personal brand identity is a "crazy difficult" question. I know many of you are sweating already just thinking about it. To ease your anxiety, our brand identity doesn't have to be a forever thing. Our brand identities can change and adapt over time. It isn't chiseled in granite with a stonemason's hammer, but it isn't written in sand, either – changing daily with the rising tide. Instead, imagine that your brand identity is carved into sandstone.

In my experience, most of us will reinvent ourselves two times – maybe three – over our career. I have reinvented myself at least three times over the past 35 years. I started out as an electrical engineer. Then, armed with an MBA, I became a management consultant with a large firm. Later on, I helped create a boutique consulting firm, which I led for 15 years. Finally, I've reinvented myself as an educator.

There's a difference between reinventing our personal brands and changing directions with the wind. As with Society's Hall Pass, we're allowed to reinvent ourselves more frequently when we're younger. As we get into mid-career, we're expected to stick with things longer and develop deeper expertise.

After a successful career, I think society gives us another hall pass when we reach the final chapter of our careers. It's not uncommon to see a banker become a nonprofit director, or an advertising executive shift to travel writing. Or perhaps an accountant becomes a college professor.

Reinventing ourselves gives each of us an opportunity to find a new spark that will take us for another decade or two of our lives. So I offer this advice: don't panic in making decisions about personal brand choices. Be thoughtful in your choices, and make sure it's something you're interested in. Work to find something you're willing to commit to for a good while. But don't lose sleep over it as if your personal brand identity were a forever thing.

References

Trevor story: Inspired by real characters known by the author. Names and locations have been changed.

Top of Mind Awareness: Paul Farris, Neil T. Bendle, Phillip E. Pfeifer, and David J. Reibstein. *Marketing Metrics: The Definitive Guide to Measuring Marketing Performance*. Upper Saddle River, NJ: Pearson Education, 2010.

Al Ries. *Focus: The Future of Your Company Depends on It*. HarperBusiness, 1996.

Al Ries and Jack Trout. *Positioning: The Battle for Your Mind*. McGraw-Hill, 1981.

Eric Gregg quote: Interview by *How to Win Client Business* research team, 2020.

Blair Enns quote: Blair Enns. *The Win Without Pitching Manifesto*. Rockbench Publishing, 2010.

7

Choosing Your Specialty

Shrink the Pond Until You're a Big Fish

> You must identify your niche, and then master it. Put in the time and effort it takes to become exceptional at a niche that generates value for clients. Mastery matters because that is what drives clients to hire you.
> —Art Gensler, architect, founder of Gensler (designer of the Apple Store)

We need a focused personal brand identity if we are to become successful rainmakers. Al Ries offers us these words of encouragement: "Focus is the art of carefully selecting your category and then working diligently in order to get categorized. It's not a trap to avoid; it's a goal to achieve. Don't let mindless criticism detract you from this goal."

For most of us, our personal brand identity doesn't appear to us all at once like a moment of inspiration on the mountaintop. It often unfolds over time – often months or years. Many successful rainmakers find their personal brand identity by connecting the dots of our resume in the rearview mirror.

Terry Pappy's Success Story: Carve a Niche, Then Carve a Niche within a Niche

Trevor's story from the previous chapter provided an example of what a brand identity *doesn't* look like: going to market doing anything for anyone. Unfortunately, there's no slot for that in the client's mind. And, if there were, it would be a poor slot to own. This reminds me of a character in John Steinbeck's novel *East of Eden*:

> Alf Nichelson was a jack-of-all-trades, carpenter, tinsmith, blacksmith, electrician, plasterer, scissors grinder, and cobbler. Alf could do anything, and as a result he was a financial failure although he worked all the time.

Terry Pappy is a talented marketing consultant, brand strategist, and host of the *Simplify & Multiply* podcast. She learned the ropes of her craft from many years in her industry including experience at ad agencies, newspapers, printing companies, and client-side companies.

If we stopped there, Terry wouldn't be that different from the tens of thousands of marketing specialists out there in the world: talented, capable, and totally unmemorable. But Terry has mastered the art of shrinking the pond. She provides marketing services to a very specific clientele: solopreneurs – one-person professional firms.

I asked Terry, "How did you arrive at your personal brand identity of serving solopreneurs?"

Terry lit up. "I didn't start out with this focus. My initial response was survival – having been laid off from my six-year span at Marriott a year after my husband passed away. All I knew was to continue delivering creative and marketing services for anyone who would hire me by the hour. Luckily, I had a few Marriott connections that caused a chain reaction of referrals that led to a Chinese menu of clients from ambulance companies, transportation companies, window installers, interior designers, hospitals, and on and on."

> Because I could do any type of marketing and creative, I became a jack of all trades taking on work simply to survive. I never looked at creating a niche. As a result, I was 100% commoditized constantly being compared to every other graphic designer, web designer, creative copywriter out there in the marketplace.

> It wasn't until years later that I looked back and saw a pattern. There was one type of client that I really enjoyed working with and got great results for: solopreneurs – individuals who operated as a "company of one." I realized I had developed a depth of understanding about the world they operated in.

That's when I decided to go all-in on serving these professionals. It was at this point that I started developing products and systems specific to that audience, which was a lot easier because I knew exactly who I was serving. And, as a result, my business really began to gain traction and grow.

Let's take a look at why Terry's brand identity is so effective. Her positioning fulfills a number of important milestones of the client's buying decision journey. For one, her identity is easy to understand and remember because it fills a unique slot in people's minds. Secondly, because of her deep experience and skill at serving this specific target audience, her professional expertise is highly respected.

Terry owns the category. When Terry speaks to solopreneurs about her marketing services, her credibility is immediately palpable. She understands the unique challenges of solopreneurs who need to build a strong brand following. Looking at Terry's brand identity from the perspective of the top client pathways, her positioning lends itself well to referrals and inquiries. Her expertise is easily remembered and respected. When a solopreneur asks around for who's the best at marketing, it's easy to see why Terry has significant top-of-mind awareness.

Bill Stoddart's Success Story: Knowing What You Believe In

Bill Stoddart is the founder of NorthFork Financial. When Bill was 40, he left his career as a high school teacher and became a financial advisor. All of the financial advisors reading this will quickly acknowledge that this is a very difficult career pivot.

Getting started in the financial advising business is difficult at any age; it takes an incredible amount of work and a lot of thick skin. And it can take many years before you see any significant success. It's especially hard mid-career when many of us have families, mortgages, and other life responsibilities.

Bill started out the same way as many new financial advisors: cold-calling by phone and going door-to-door – not everyone's idea of a fun job. Bill drove to neighboring towns to knock on doors because he was embarrassed to do so in his own community. He was afraid a former student or parent would recognize him from his years at the local high school.

Bill acknowledges that this cold prospecting wasn't very effective. No surprise to us, really, when we look at the relatively low success rate of cold prospecting in the client pathways data. Fortunately, Bill was on a small salary with the regional broker-dealer firm he was working for at the time. But this income would not last for long. He knew he needed to find a new approach to winning client business.

To understand the turn that finally led to Bill's ultimate success, we need to go back a few years to when his father passed away. Bill was still in his 20s at the time. His father left him $25,000 in his will, and Bill was shocked to find that the money was invested in tobacco stocks.

Bill had a strong negative reaction when he learned his dad's money was invested in tobacco companies. He couldn't conceive of making money off of selling products that would make people sick. It seemed very wrong. Bill sold the tobacco stocks and reinvested the money in socially responsible companies, which was then a very new thing in the early '90s.

Fast-forward a few years and Bill found his niche in helping clients with a similar point of view regarding investing – specifically, clients who had a desire to invest in socially responsible companies. Once Bill landed on this path, his business began to grow. The timing of Bill's new brand identity coincided with a growing movement of investors who wanted to attain financial security while feeling good about the companies they invested in. Bill describes his investment philosophy as "where money and meaning come together."

Looking at Bill's success through the lens of the client's buying decision journey, it's easy to see why his brand identity has been so effective. His investment philosophy is clear and stands out. And his personal story and investment track record are highly credible. From the perspective of the top client pathways, Bill's personal brand identity lends itself to referrals and inquiries. Furthermore, his clear target audience of socially responsible investors allowed him to find those who had similar values and respected his approach.

Three Ways of Defining Your Personal Brand Identity: Functional Expertise, Target Audience, and Geographic Focus

Terry's and Bill's stories apply to those of us working in larger firms as well. Within larger firms, it's equally important to have a strong brand identity. Firms are comprised of individuals working together toward a common goal. The stronger and clearer our identity for being good at a specific thing, the more valuable we will be to the firm. And, subsequently, the more referrals you'll get from the firm's partners and clients.

Whether working as a solo practitioner, a partner in a boutique firm, or for a large services company, there are three primary ways to define your personal brand identity:

- Functional expertise
- Target audience
- Geographic focus

Often, your personal brand identity will be a unique combination of these three variables. Let's unpack these to see how each helps define how others see us.

Functional Expertise

The first approach in defining our personal brand identity is through a specific functional expertise. If you're an attorney, for example, you may specialize in estate planning. Or you may be an expert at business litigation or real estate transactions. In the western U.S., where I live, water rights are a big issue for farmers, ranchers, and municipalities. Subsequently, there are attorneys who specialize in the state and federal laws relating to this often scarce resource.

The same holds true in other professions as well. Some architects specialize in residential homes. Within residential, some architects build an expertise around low-income housing; others focus on luxury homes. Other architects specialize in commercial buildings; some focus on schools and others on hospitals.

Within management consulting lies a wide variety of functional expertise, including HR, supply chain logistics, strategy, and organizational design. If you're a solo practitioner, you can't be an expert in all of these areas. In order to be respected, you'll have to develop an expertise in one area. Often, your specialty will be a niche within a niche. If you work for an IT consulting firm, you may specialize in cyber security or data storage in the cloud.

Target Audience

The second approach in defining our personal brand is through our target audience. Specific target audiences often have unique needs that benefit from having deep knowledge. You see this in every type of service profession.

In the executive search field where companies hire outside experts to fill high-level positions, you will typically see individuals serve a specific industry – say, high-tech or pharmaceutical companies. In the environmental services field, you will see individuals and boutique firms specialize in serving the petrochemical industry, or municipal wastewater treatment. The same holds true in the legal profession, where you will find attorneys specializing in business franchises or maritime shipping.

Among registered dieticians, some specialize in the needs of premature babies, while others concentrate on the specific needs of the elderly, diabetic patients, or professional athletes. In investment banking, you'll find those specializing in the capital needs of mid-cap companies, and others with state and local governments.

Or you may serve a niche psychographic market, as we saw in Bill Stoddart's story – individuals having a specific set of beliefs and values. Many professionals serve very specific demographic niches: high-income, low-income, minorities, women, Native American, and so on. Often the choice of target audience is deeply personal, reflecting your own personal beliefs.

Geographic Focus

The third primary way in which professionals specialize is through geographic focus. Each geographic market has specific needs tied to their distinct legalities, practices, and cultures. In the legal profession, each U.S. state has a unique legal structure. Some laws in California may be very different from the laws in Texas.

The same is true in accounting. Each state has its own tax code that is unique to the people living and operating there. In some states, there is no state income tax (Alaska, Florida, Nevada, South Dakota, Texas, Washington, and Wyoming), and other states have a myriad of local, county, and state taxes to contend with.

Some professionals focus on a small, local market while others have a regional, national, or global focus. The depth of your expertise often influences the breadth of your geographic focus. You may serve a national architectural market if you specialize in LEED-certified commercial buildings (Leadership in Energy and Environmental Design). If you specialize in middle-income residential homes, you may serve just one or two counties.

Due to the density of their target audience, some professions lend themselves to local or regional markets, while others lend themselves to national or global markets. Personally, I would go hungry if I focused on helping those just in my hometown or state. My city's population simply isn't large enough (roughly 50,000 residents) to support the work that I do, nor is the population of my state (roughly 1 million). In order for me to find the people I wish to serve, my market must be national if I'm going to have a sufficient market size for my specific expertise.

Finding Your Unique Combination of These Three

Establishing your unique brand identity is a combination of these three main variables: functional expertise, target market, and geographic focus. The process of finding your unique combination is a process of trial and error and often evolves over time. (See Figure 7.1.)

The first variable to choose is your functional expertise. Your education, accreditations, and work experience will help identify your functional expertise. Many of us find our functional expertise through the work that we do in our careers long before we're tasked with becoming a rainmaker.

Through these experiences you may find a particular area within your broader functional category that particularly interests you. When you find an area that excites you, it's a good sign this is an area to go deeper in. Over time, you'll build

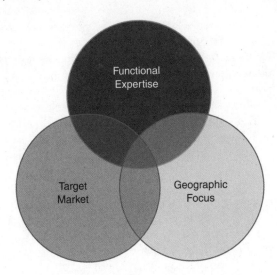

FIGURE 7.1   Three Ways of Defining Your Personal Brand Identity

more and more expertise in an area that could become your go-to expertise that you wish to be known for.

What if you're uncertain about going deep in a particular area? Or what if you have more than one area that interests you? One idea that may be helpful is what I call the *five-year rule*, which is simply this: *Would you be willing to dedicate five years of your career to becoming an expert in this particular area?*

What is special about five years? For one, five years is a long enough period of time in most functional areas to become reasonably expert. Clearly, this isn't true of all specialty areas (say, becoming a neurosurgeon). But, for many professions, once you have the foundational knowledge in place, five years is sufficient to give you expertise and credibility. Five years also ties in nicely with the "10,000 hours of practice" to become an expert popularized in Malcolm Gladwell's book *Outliers: The Story of Success*. The average work year is 2,000 hours, and five years puts you in the range of becoming an expert in many areas.

Additionally, five years is sufficiently short to remove the panic of choosing one path for life. If choosing a career niche is hard, it would be paralyzing if we felt our decision was forever. In keeping the decision to a five-year span, it helps make the decision less daunting.

The second variable to build your personal brand around is your target audience. Choosing your target audience may be shaped by your work experience long before the need to demonstrate your rainmaking skills. As with your functional expertise, your experience with a specific target audience often comes from your work experience.

Many professional services firms are organized along industry lines – and our early career experiences will expose us to a variety of target audiences. If there is one industry focus that particularly interests you, again, this is a good indication of an area for further specialization.

Choosing your target audience can also be based upon more personal reasons, as saw in Bill Stoddart's example.

The last primary variable defining your personal brand is geographic focus. As we discussed, the breadth of the geography you serve is partially a result of the depth of your expertise. Generally speaking, the deeper and more niched your service, the wider you'll have to cast your net to gather a client base. This is largely a result of finding a large enough market for your services, as is the case with my practice.

Another consideration when choosing your geographic focus ties back to TOMA (top of mind awareness, discussed in Chapter 6). You have to be among the top three providers for what you do in order to find consistent success. What do you do if you're not in the top three? You guessed it: shrink the pond.

If you're not in the top three in what you do in the U.S., shrink the pond. Not in the top three in the region of the country where you operate? Narrow your focus to the state level. Not in the top three in your state? Narrow it again to the city where you work. Over time, as you build your brand reputation in a region, you can begin to expand your geographic focus to serve a larger community.

References

Terry Pappy story: Interview by *How to Win Client Business* research team, 2020.
Bill Stoddart story: Interview by *How to Win Client Business* research team, 2020.
Malcolm Gladwell. *Outliers: The Story of Success*. Back Bay Books, 2008.

CHAPTER

8

You Can't Sell Beyond Your Credibility Zone

The Cautionary Tale of EDS

There is one fundamental I can't teach you – you have to have a bona fide expertise. There's no faking expert knowledge.
—Lee W. Frederiksen, Elizabeth Harr, Sylvia S. Montgomery, *The Visible Expert*

Creating a strong personal brand identity is the first of the five rainmaker skills for a reason: without one, we won't have a seat at the table when our client makes a decision. And we won't be top of mind when a prospective client asks a colleague for a referral. An unfocused, undifferentiated brand identity is like the bland decor we find in every government office.

While not an easy process or an exact science, you've got a good start at understanding how to go about defining your personal brand identity through the careful selection of your functional expertise, target audience, and geographic focus.

Before we transition to the second rainmaker skill – *demonstrating your professional expertise* – I want to share with you a cautionary tale to illustrate what happens to those who stray from a strong brand identity. It's a personal story. It's part of my life's history. It's the story of EDS (Electronic Data Systems).

There are two common mistakes I've witnessed as it relates to flaws in an individual's – or a firm's – brand identity:

1. **Not having a clear brand identity:** Trying to be all things to all people
2. **Straying from a strong brand identity:** Having a clear brand identity and expanding into areas beyond your credibility zone

The EDS story falls into the second category. While this is a story about a company, the lessons apply to us as individuals as well.

The Story's Beginning

I vividly remember the day. It was in late March 1993, during my first year of graduate business school. I stood in our small kitchen as I opened the envelope. The letter was from Arthur D. Little, a management consulting firm based in Cambridge, Massachusetts. This letter was either an offer for a coveted summer internship or a short, sweet rejection.

As I took a deep breath, hands shaking slightly – I opened the letter. The first sentence began, "We are pleased to offer you a position in our summer internship program…" I exhaled as waves of relief washed over me. Then the partying began. The rest of that day is a bit hazy.

Most of us have similar memories that serve as milestones in our personal journeys. I knew this letter signaled an important moment in my career. My decision to leave GE and pursue my MBA was beginning to pay off. The starting salary was double what I was previously making before grad school.

I had a wonderful summer internship with Arthur D. Little. As it turned out, my car headed west to L.A. rather than north to Boston. The firm had an opening in their West Coast office, and I jumped at the chance to experience Southern California. Don Scales was my boss that summer – the VP in charge of the West Coast office.

Later that fall, upon arriving back in Charlottesville, I got a call from Don. I was studying in the living room when the kitchen phone rang. (This was in the days before cell phones.) Don explained that he and his entire team were now working for EDS. He hoped that I would join his team after spring graduation.

I now had two offers of employment waiting for me after school: one from Arthur D. Little and the new one from EDS. Looking back I don't remember it being a difficult decision. My colleagues I had worked with at Arthur D. Little were excited about their new employer. It didn't hurt that EDS bumped my starting salary by $10,000 – a lot of money to a 28-year-old with student loan payments.

The Story's Middle

EDS was a household name in the mid-'90s. It was the brainchild of the wacky billionaire presidential candidate Ross Perot, famous for his charts and graphs on the national debt. Perot didn't win the 1992 presidential election, but his company continued to thrive in an era of expanding corporate and government IT investment.

By the mid-'90s, EDS was a $10B company with over 50,000 employees worldwide. In 1994 they signed a deal worth $3.2 billion with Xerox, which at the time was the largest information technology contract ever. As often happens to companies, EDS's grand success led it astray: the firm began to believe that it could do anything. If it could sell billion-dollar IT contracts, how hard could it be to sell million-dollar projects in management consulting?

The logic of EDS's attraction to management consulting was easy to follow. While the profit margins on IT services might be, say, 10%, the margins on consulting services were often double or triple that. EDS wanted a piece of the action.

At the time, I didn't know much about management consulting; I had spent a total of three months in the profession. I was, however, mesmerized by EDS's glamourous glass and steel headquarters built 20 miles north of Dallas. And I bought into the belief that EDS Management Consulting would quickly become a competitive threat to the other top consulting firms.

As it turned out, millions of dollars of investment in new employees, office buildings, and advertising could not buy EDS the respect of Fortune 500 CEOs. It soon discovered that transferring its great reputation in IT services to the field of management consulting was a bridge too far in the minds of corporate executives. EDS Management Consulting was burning through countless millions and the top brass were losing patience. They were in search of a new strategy for their struggling management consulting division.

In 1995, EDS changed course. If it couldn't work its way into the front door of corporate America, it would try a new angle – maybe the side door was open. EDS acquired the reputable Chicago-based consultancy A.T. Kearney, then the fourth-largest independent management consulting firm in the world whose top clients included GM and Coca-Cola. According to an article in the *New York Times*, the acquisition was for $600 million. A.T. Kearney's 118 partners were instant millionaires.

I can still remember the day EDS made the acquisition announcement. My teammates and I were very excited – and also curious what this meant for us. Would A.T. Kearney accept us into the fold? How would the cultures of EDS and A.T. Kearney fit together? What would clients think of this newly formed company?

The people at each company were very different. EDS was a techy, Men's Wearhouse crowd – a holdover from Ross's days at IBM. A.T. Kearney was more Brooks Brothers – not unlike what you might expect to find on a leafy Ivy League campus. Both were accustomed to being the smartest in the room.

The Story's Ending

The merger didn't go well. In retrospect, the story should be an MBA case study, if it isn't already. Melding two very contrasting cultures – both accustomed to getting their own way – is problematic if not outright impossible. As a result of the merger, the reputations of two well-regarded companies were both tarnished. EDS had lost focus in pursuit of growth. A.T. Kearney lost its reputation for independent thinking and analysis.

The ten-year struggle included many leadership changes, layoffs, pay cuts, and a relocation of A.T. Kearney's HQ from Chicago to Dallas. Many of A.T. Kearney's best and brightest were long gone. In the end, EDS sold A.T. Kearney back to its partners for an undisclosed amount – presumably a lot less than the $600M it paid to acquire it.

A.T. Kearney has found its footing since regaining its independence in 2005. Today it is rebranded as Kearney, with 60 offices in 40 countries. It is once again regarded as one of the top management consulting firms in the world.

The rest of the EDS story is a mixed bag of good and bad – mostly bad. It was subsequently sold to HP in 2009. HP struggled to find its balance in the shifting sands of hardware and software services. Today, after a series of mergers and spinoffs, the once venerable EDS brand no longer remains.

The Lessons of the Story

I worked for EDS and subsequently A.T. Kearney for a total of two years – hardly a long career with either company. But I did manage to learn a number of important business lessons.

The biggest takeaways from my experiences at EDS and A.T. Kearney are:

- Takeaway 1: The importance of having a strong brand identity
- Takeaway 2: The futility of trying to sell beyond your credibility zone

Without a strong brand identity, clients struggle to find a mental slot for us to fit into. Second, trying to sell services outside of our credibility zone is an expensive waste of time. The only clear path to success is to carve out a niche that you can own, and to stay true to that niche. You can expand your niche over time, but it has to be a gradual process of expanding into closely related niches. Jumping into new niches far removed from your own leaves clients wondering, "Huh?"

These lessons have guided my business thinking for the past 25 years. And the lessons learned from these two companies apply to our personal brand identities as well, whether you are an aspiring partner at a large firm or striking out on your own.

A Seat at the Table

I'll leave you with one final thought as we move on to the second rainmaker skill. It is a thought that was shared with me recently by Jacob Parks, COO and partner at Profitable Ideas Exchange (PIE), an industry leader in building executive round-tables focused on the highest-priority business topics of the day.

Jacob offered this perspective:

> When you look at most websites in any profession, they all look the same – they say the same thing. They are virtually indistinguishable. Firms – as well as individuals – need to stay true to who they are. They need to ask themselves: *Where do we have a right to a seat at the table?*

I like that saying – where do we have a right to a seat at the table?

As it relates to choices about your personal brand identity, where does your resume give you a right to a seat at the table? The answer to that one question may be the best guiding star as you chart your path to a strong personal brand identity.

The Best Advertising Mind You've Never Heard Of

Rosser Reeves is no longer a name most of us are familiar with. He has long since passed away. But his little gem of a book, *Reality in Advertising*, was hugely influential when it hit the bookshelves in 1961. It remains a classic to this day in the field of advertising and is still taught at Harvard Business School.

Born in rural Virginia as the son of a Methodist minister, Rosser briefly attended the University of Virginia before being expelled for crashing his friend's car after a wild night of drinking. Rosser found himself in Richmond working in a bank. As today, banking was a solid choice for a young man to get started in business – and back then you didn't need a college degree to get your foot in the door.

Rosser proved to be a bad fit for the numbers side of business, but he quickly demonstrated his gift with a pen. This landed him a job in New York working with the advertising firm Ted Bates & Co. He went on to have a successful career in advertising, and eventually rose to the position of chairman of the board at Ted Bates & Co. His long list of memorable advertising campaigns includes the successful presidential campaign of Dwight D. Eisenhower in 1952.

Rosser's biggest contribution to the field was an advertising concept called Unique Selling Proposition (USP). Rosser put the USP concept to use with his clients, helping grow Ted Bates & Co. from a $4 million to a $150 million agency.

Rosser believed that a product's USP needed three things:

1. Each advertisement must make a proposition to the consumer: "Buy this product, and you will get this benefit."
2. The proposition must be one that the competition either cannot, or does not, offer. It must be unique.
3. The proposition must be so strong that it can pull new customers over to your product.

Rosser's USP may have originated in the advertising world of consumer products, but I think it has value to us in professional services. Having a unique selling proposition can give us an edge in helping strengthen our personal brand identity. It may very well be the secret sauce that differentiates us in the minds of those we wish to serve.

What's unique about your brand identity? It could be a breakthrough project with a high-profile client. Or depth of expertise over decades that others can't easily replicate. Or perhaps a methodology or point of view that is different than competitors. Whatever yours is, having a USP is a proven way to remain top of mind when clients are searching for expert help.

References

EDS story: Inspired by author's experiences, 1992–1996.

Allen R. Myerson. "EDS to Buy Consultant for $600 Million." *New York Times.* June 7, 1995.

Vivek Velamuri. "EDS's merger with A.T. Kearney." *Financial Times.* July 1, 2013.

"Our Story." Kearney website: https://www.kearney.com/about-kearney/our-story.

Rosser Reeves. *Reality in Advertising.* Alfred A. Knopf, 1961.

Skill 2: Demonstrate Your Professional Expertise

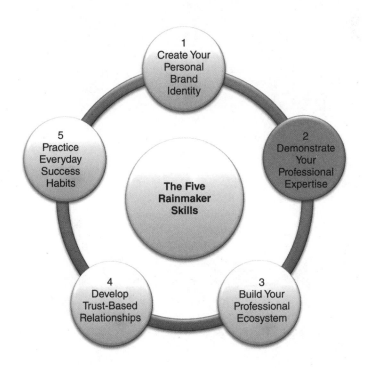

9

How Clients Tell Who the Real Experts Are

Clients Need Clues That We Are Really Good at What We Do

The fact that nearly everyone of importance in Wall Street could be found at the Waldorf made it a highly revealing laboratory for the study of human nature. The various "rooms" of the Waldorf – the Empire Room or Peacock Alley, the billiard room, and the Men's Café with its famous four-sided mahogany bar – were exhibition galleries in which every human trait was on display. Sitting in these rooms, it was always an intriguing exercise to try to detect the doers from the braggarts, the genuine human article from the spurious.

—Bernard Mannes Baruch, America financier and presidential advisor

You may recall our simple thought experiment from the book's introduction:

Why is it that we have no problem whatsoever buying a home after an hour-long walk-through, yet will agonize for months in choosing an architect if building a custom home?

One of the key reasons is that prospective clients have a difficult time in identifying the true experts. Rainmakers are skilled at providing prospects with clues to their authentic expertise, signs that we are really good at what we do. I think of these clues as channel markers – the buoys that safely guide a boat from sea to harbor while avoiding hidden dangers.

In providing channel markers, we assist our prospective clients in avoiding the rocks and sandbars hidden in their decision-making journey. And when we do this effectively, we increase the likelihood that we'll win the client's business.

Clients Seek to Avoid Regret

Most of us have never heard of Amos Tversky and Daniel Kahneman; if we have, we probably learned of them through Michael Lewis's fascinating book *The Undoing Project*. Lewis is famous for taking wonky, esoteric topics and making them accessible to broader audiences. He is well-known for his best-selling books *Moneyball*, *The Big Short*, and *Liar's Poker*. *The Undoing Project* is an interesting story of the complex relationship between Tversky and Kahneman, and a look at how their important research shaped our understanding of how humans make decisions.

Tversky and Kahneman first met as young PhD psychologists in the 1960s at Hebrew University in Jerusalem. Their work together over the coming three decades led to many breakthrough discoveries in the workings of the human mind, particularly as it relates to understanding the biases we hold when faced with decisions.

Tversky ultimately ended up at Stanford and Kahneman at Princeton. Kahneman was awarded a Nobel Prize in Economics in 2002, and Tversky surely would have as well had he not died of cancer in 1996 – Nobel prizes are not awarded posthumously.

One of the hallmarks of their research called into question the idea that humans are rational decision makers, as the field of economics had claimed for over 200 years. Their research proved that as humans we are often misguided by biases that cause us to make decisions that are not altogether rational, at least not in the way economists once thought. As humans, we are much more complicated than traditional economic theory would have us believe.

Tversky and Kahneman contributed heavily to the creation of a new field called behavioral economics, which studies the effects of psychological, emotional, and cultural factors on the decisions of individuals and institutions, and how those decisions vary from those implied by classical economic theory.

One of the early topics that fascinated Tversky and Kahneman was the concept of regret as it related to its influence on decision-making. In a memo to Tversky in the 1970s, Kahneman wrote:

> It is the anticipation of regret that affects decisions, along with the anticipation of other consequences. When people make decisions, they do not seek to maximize utility. They seek to minimize regret.

Each of us has felt regret many times before, ranging from regret about little decisions (I wish I had purchased the other pair of shoes) to bigger decisions like choosing a college major.

When making buying decisions about products, say a house or car, it's easier for humans to minimize the likelihood of future regret because our satisfaction is easier to gauge. Conversely, it's harder for us to assess the likelihood of future regret when deciding on an expert to assist us in solving an important problem. As regret-avoiding creatures, humans seek out information to help minimize this uncomfortable feeling.

Because human decision-making is influenced by regret avoidance, any steps we can take as professionals to reduce this anticipated feeling will increase our chances of success. In much the same way that channel markers guide a ship's captain safely to harbor, this rainmaker skill leads clients to a good decision while minimizing anticipated regret.

How We Can Help Clients Avoid Anticipated Regret

Clients face a myriad of risks when choosing a professional. These risks include:

- **Competence risk**: Is the professional really good at what they do?
- **Culture risk:** Is the professional a good fit culturally? Do they share similar values with the client?
- **Performance risk**: Will the professional actually follow through on doing what they say they will do?
- **Integrity risk**: Will the professional's motives be pure? Will they do what's best for the client at all times?
- **Reputational risk**: How will this hurt the client's reputation if the project ends poorly?

- **Financial risk**: How badly will this impact the client's financial performance if things don't go well?
- **Career risk**: Will the client's career be derailed if this project goes badly? Will they be blamed for this?

With the help of Tversky and Kahneman, I think we can see these client risks in a new light. Regret avoidance is really just another way of thinking about risk mitigation. If we can help clients avoid the feeling of anticipated regret, we are helping them reduce the risks of a bad outcome associated with their buying decision.

There are many techniques for demonstrating our expertise. We can – and should – select the methods that are best suited for us. The ways in which we choose to help clients see that we are true experts are simply different approaches to achieving the same outcome: helping clients avoid the feeling of anticipated regret.

We are acting as valuable guides in the decision-making journey when we help prospective clients minimize anticipated regret. We understand how they feel and empathize with their situation, and help them navigate through the process in a way that is not manipulative or dishonest. And we feel better about our role as a facilitator in their journey, because we abhor the idea of being seen as a salesperson.

Unlearning the Mental Hang-up We Have with Talking about Ourselves

If we are to become successful at mastering the skill of demonstrating our professional expertise, we must unlearn this widely held belief: *the perception that promoting our expertise is bragging*. Just as we resist the notion of ourselves as salespeople, we equally resist the idea of seeing ourselves as braggarts. I can still hear the kind admonishment of Ms. Coby, my first kindergarten teacher, "Don't brag, Doug. Bragging isn't nice and no one likes a braggart."

Mindset Shift Two: Viewing our professional accomplishments as credibility markers rather than bragging

There's a big difference between bragging and demonstrating our expertise. And, independent of Ms. Coby's good advice, we have to overcome this misperception if we are to become successful at demonstrating our expertise.

If clients seek to avoid the nagging feeling of regret in making a bad decision, any steps we can take to help avoid this experience is welcomed. Viewed in this way, demonstrating our expertise becomes a helpful service that we can provide. This is vastly different than the widely held belief that talking about ourselves is bragging.

The art is in learning to talk about ourselves in a way that isn't bragging. When we stick to facts and avoid bold, unfounded assertions, we demonstrate our capabilities in honest and authentic ways. When we share our credibility markers with those we wish to serve, we aren't bragging. We are simply helping prospective clients in their decision-making journey. We are helping others see our authentic skills so that they can avoid the feeling of regret.

Let's conduct another quick thought experiment, this one involving a prospective client, Raymond, who needs help in designing his new website. Raymond has narrowed his search down to two web developers: Developer One and Developer Two. Suppose that both of his options are equally honest, hard-working, dependable, and capable – and priced roughly the same.

However, with an intriguing twist to the story, suppose that Developer Two has done a better job of demonstrating her expertise than Developer One. Developer Two has highlighted recent work on her website, has recently spoken on the latest digital trends at a business conference, and teaches digital marketing at a local university. Developer One has done none of these.

Developer One	Developer Two
Honest, hard-working	Honest, hard-working
Experienced and capable	Experienced and capable; case studies on website
	Speaks at digital conference
	Teaches at local university

Who do you think Raymond will choose: Developer One or Developer Two? All other things being equal, he will likely choose Developer Two. In his decision, Raymond feels less risk – less anticipated regret – in choosing Developer Two. There were more clues provided about Developer Two's capability.

And because Raymond isn't an expert in web development, he is grasping for clues as to who is better. Although Developer One has just as much relevant experience for this work, he provides no credibility markers to help guide Raymond through his decision-making journey.

Who is better off in this scenario? Clearly, Developer Two – the winner. But is Raymond better off? Hard to say – what if in fact Developer One would have been the better choice? In not demonstrating his expertise more effectively, Developer One seemed the riskier choice.

- Did Developer One get hung up on the notion that demonstrating his expertise is bragging? If so, perhaps he didn't offer valuable examples that would have assisted the client in making a better decision.

- Perhaps Developer One didn't fully appreciate how difficult this decision is for prospective clients. Maybe he thought, "Clearly he can see I'm the more talented and experienced developer." Because of information asymmetry, Raymond looked for signs as to who was the more capable developer.
- Alternatively, maybe Developer One just didn't know how to demonstrate his expertise in a tasteful way. Maybe he simply didn't know how to show clients how talented he was without feeling like a braggart. Perhaps he would gladly provide these clues if only he knew how to do so.

Armed with this knowledge, we're not going to make the same mistakes as Developer One. As aspiring rainmakers:

1. We understand how difficult the buying decision can be for clients – the journey is difficult, often confusing, and filled with perceived risks and anticipated regret.
2. We know that clients need clues to our expertise and experience.
3. We know the importance of providing credibility markers to demonstrate out capability – because clients can't see inside our heads.

We're ready to learn how to demonstrate our expertise in a professional manner. To show others how talented we are in ways that aren't boastful. To learn to become a skilled rainmaker in a way that feels good to us.

Choosing Your Credibility Markers

There are a variety of ways in which we can demonstrate our expertise to prospective clients. The choices you make in how you demonstrate your expertise will depend upon a number of factors. One factor is your profession. What may work in the legal profession may not be ideal for a marketing specialist; what is highly effective in strategy consulting may not be that helpful in financial services.

Another factor is what fits best with your personal skills and preferences. Because all of these techniques are simply tools that achieve the same outcomes, we are free to choose the tools we like best and are best at doing.

Independent of your personal choices on which credibility markers to use, there are two vital principles that must guide the way:

- **Principle 1: Honesty.** We must be honest in communicating our expertise to the world.
- **Principle 2: Consistency.** We must be consistent in sticking with the approaches that work best for us.

In everything we do as professionals, honesty wins. There is no such thing as a dishonest trusted advisor – except maybe in *The Sopranos*. There is no room for making claims or saying things about ourselves that aren't true. As the famous Confucian saying goes, "Three things that cannot be long hidden: the sun, the moon, and the truth."

When we're honest about our skills, expertise, and experience, sharing these with those we wish to serve is highly appropriate. We're a valuable partner in their decision-making journey. Making claims that aren't true is a sure way to ruin our reputation.

Consistency in the approaches we use to demonstrate our expertise contributes to our personal brand identity. It becomes a part of who we are. Over time, persistent use of specific credibility markers helps define us and helps others remember us. Consistency in our approach helps fill an empty slot in the client's mind and creates top of mind awareness.

The good news is you have lots of latitude in how you demonstrate your professional credibility. Your choices will be influenced by your profession and your unique skillset. Once you've landed on an approach or two that works for you, you'll be guided by the time-tested principals of honesty and consistency.

References

Bernard Baruch quote: Bernard Baruch. *My Own Story*. Henry Holt and Company, 1957.

Amos Tversky and Daniel Kahneman story: Michael Lewis. The Undoing Project: *A Friendship That Changed Our Minds*. W.W. Norton & Company, 2017.

Daniel Kahneman. *Thinking, Fast and Slow*. Farrar, Straus and Giroux, 2011.

Herbert A. Simon. *Models of Thought*. Yale University Press, 1979.

CHAPTER

10

How to Toot Your Own Horn without Looking Like a Jerk

Proven Techniques for Demonstrating Your Expertise

This is not an exercise in egotism. In a funny way, this is not even about you. Rather, it is about establishing relationships within an ecosystem for the benefit of all participating.

—Geoffrey A. Moore, author, *Crossing the Chasm*

Demonstrating your expertise requires a commitment of time and energy. I frequently hear, "Who has time for this?" Because clients often struggle to discern who the true experts are, the time we invest in demonstrating our expertise will increase the odds of our success.

It is much easier to commit to doing something that we actually enjoy. So, don't start something that seems like a huge burden. If you like it, you'll find time for it. It may be nights and weekends at first, but in time you'll carve out a schedule that feels more manageable. If you're like many, you will actually grow to look forward to these activities.

Over time your efforts may actually help reduce the effort spent on other client development activities. If you do an effective job of demonstrating your expertise, perhaps you'll spend less time defending your *bona fides* with new clients. Or, with an increase in referrals and inquiries, you'll spend less time prospecting for new clients.

While creativity in many aspects of our work is highly valuable, in choosing how best to demonstrate our expertise, it's advisable to follow a handful of well-worn paths. There are enough proven options to choose from that gives us ample flexibility in aligning with our skills and preferences.

Technique 1: Writing

Noel Sobelman is a principal at Change Logic, an innovation consulting firm that helps clients ideate, incubate, and scale new businesses. For two decades, Noel has shared his perspectives on his industry through his writing. He has written dozens of articles and white papers, and these insightful publications have proven to have a surprisingly long shelf life. Noel enjoys writing. I've never met a writer who doesn't enjoy writing.

As a former engineer, Noel admits that his papers tend to get a bit detailed and longish. And his papers can take him a while to write. Due to the depth of his papers, Noel says he used to get out about two to three papers per year on average. However, with consistency and persistence over the last few years, he has accelerated his output and now has about 20 articles posted to his LinkedIn profile.

Noel has discovered that his papers have remained relevant long after they were written. In some cases, he's able to take an older paper – one that he wrote ten years ago, for example – and refresh it and post it again. One of Noel's newest clients discovered him online through one of his published white papers.

Noel said something that struck me: he only writes about topics that he's passionate about. In writing, he's able to research a topic of personal interest and it helps him stay current in his field. Given the time commitment that good writing requires, his interest in each topic helps him see each paper through to completion.

Writing is one of the most effective ways of demonstrating our expertise. When we write about a topic, it communicates our professional credibility to clients. Your knowledge might be in your head, but clients can't read our minds. Putting our thoughts in writing helps us convey our knowledge and can signal that we are really good at what we do.

Writing can take many forms, from white papers to articles published in a trade magazine, from blogs to published books. Many agree that getting an article published in the *Harvard Business Review* carries more weight than a self-published white paper. And published books may signal credibility in a way that a blog post doesn't. I would say don't worry about where your work gets published; just start writing.

Your writing will get better with practice. Maybe you start with a quarterly newsletter, or a monthly blog sharing your thoughts on industry best practices. Over time, like Noel, your consistent efforts will pay off in demonstrating your professional expertise to those you wish to serve.

Technique 2: Public Speaking

Amir Tohid is the co-founder and managing partner of Analytics Effect, Inc., based in the Philippines. Amir's company specializes in digital analytics and online measurement. He speaks regularly to business groups in Manilla, sharing his knowledge of digital analytics and digital marketing trends.

Amir has become a magnet for people seeking help in promoting their businesses online. Whenever he speaks, people always hang around to chat with him further about where he sees the industry going. Amir discovered public speaking opportunities by attending local business happy hours. These events often include a presentation on a current business topic. Amir sought out the event planners and offered to speak on digital analytics and marketing – a topic relevant to many businesspeople.

Amir observed that speaking was an effective way for him to connect with those he wanted to get to know in his business community. After each talk, he shared his slide deck with anyone who was interested. Over time, his growing reputation as a thought leader led to more speaking invitations.

It is interesting that a guy specializing in digital marketing found public speaking to be his preferred approach. Intuitively, you would think that a digital expert would prefer an online medium. Amir's response intrigued me: "Our work may be digital, but it starts with a relationship. Getting face to face with people helped me built trust and professional respect. And it gave me a platform for being perceived as an expert in my field."

Public speaking has the benefit of helping you connect with your target audience and establish your reputation as a go-to expert on a topic. When we speak in public, it allows us the chance to share our expertise with a community of interested people. In public speaking, we go from being a member of a community to being a thought leader in the community.

While public speaking doesn't appeal to everyone, it can be a highly effective way of building our professional reputations. If you like speaking, then start small and get started. Speak at local meetings or small regional conferences. With practice, your speaking skills will improve and your opportunities will increase. Along the way, you'll build your personal brand reputation and create connections with those you wish to serve.

Technique 3: University Teaching

Billy Newsome is a tax attorney in Columbia, South Carolina, and the founder of Newsome Law. Billy specializes in tax, estate planning, and trust litigation for high-net-worth individuals and families. He has been practicing law for over 25 years in the Southeast.

In addition to his law practice, Billy also teaches estate planning at the University of South Carolina School of Law. Through his teaching, Billy is able to share his knowledge with the next generation of attorneys, and stay connected to the law school's faculty and alumni. Teaching at the law school also has the benefit of helping him stay sharp in his field. Teaching provides the sneaky benefit of forcing you to know a topic in greater depth than you did before.

Billy also believes that teaching is a form of community service – a way of giving back. Because the law school provided him many opportunities in his career, his commitment to teaching is a way of saying thanks, and paying it forward to others. He quickly acknowledged that it requires a time commitment.

But there are professional benefits in teaching that go beyond the psychic reward of helping others. Professors are respected members of society. By adding law professor to his resume, Billy has increased his reputation through the platform that the position provides. It's one additional credibility marker that signals to prospective clients that he is an expert at his craft.

Many colleges need people like Billy to fill out their faculty roster. Whether in law, business, architecture, or engineering, universities hire a significant number of nontenured instructors to teach their classes. Instructors like Billy help bring practical, real-world experiences to the classroom.

Like writing and public speaking, teaching at a university is a proven credibility marker. If you think you might like to give it a try, reach out to your local university. It's a great opportunity for you to give back to your community, to stay current in your field, and to build your reputation as a leader in your profession. If you don't live in a community with a local university, there may be online teaching opportunities. A colleague of mine teaches a business course at a prestigious university 2,000 miles from his home.

Technique 4: Radio Programs and Podcasts

Mike McCormick is the founder of McCormick Financial Advisers, an independent adviser helping individuals and families plan their financial futures. Mike is also the host of a weekly radio program and podcast called *Preserving Your Wealth*.

Each week Mike shares his knowledge of what's happening in the economy and financial markets with local radio listeners and with a national audience via his podcasts. His programs offer listeners a chance to call in and ask questions relating to their personal situation. Mike's deep knowledge of finance and his sharp wit make for a lively program.

I marveled at how Mike made hosting his programs seem so effortless. "Ha!" he laughed. "It might look easy, but it took many years." Like many others, Mike recognized that hosting his weekly program was a significant time commitment. But he quickly added that each year he gets better and more efficient.

Radio – and increasingly podcasts – are proven platforms for demonstrating professional expertise. Radio talk shows have been around for decades – especially for business, finance, and politics. Podcasts are a relative newcomer, having been around only since the 2000s, but have exploded in the past few years.

I've long been a fan of business talk radio and subscribe to SiriusXM. One of my favorite programs, *Marketing Matters*, is hosted on Wharton Business Radio by Professors Americus Reed and Barbara Kahn. Each week they discuss marketing and advertising trends, consumer behavior, brand-building, and much more with their expert guests.

I'm sure Americus and Barbara invest a considerable amount of time in their weekly show. With all of their other teaching and research commitments, why would they go to all the trouble? Because hosting a radio program is a proven approach to building a reputation as a subject matter expert.

While I'm certain that Mike, Americus, and Barbara make it look easier than it really is, DIY technology is making it easier every year to host a radio program or podcast. If you're inclined to try this approach, for a relatively nominal investment your home office could become a recording studio.

Like many of the other credibility-enhancing techniques, getting started requires a time commitment. But, as with writing, speaking, and teaching, hosting a radio program or podcast can be an effective way of connecting with your audience and building your personal reputation as a thought leader in your field.

Technique 5: Serving on a Board of Directors

I first met Merilee Glover about 20 years ago when I joined the board of directors of our local chapter of Big Brothers Big Sisters. Merrilee served as the board's treasurer. In her day job, Merrilee was a partner at a large regional accounting firm.

Each month at our regular board meetings, Merrilee carefully explained the nonprofit's financial position to the board. It was very clear that she had done her homework, and patiently answered questions from board members.

When my business partners and I were looking for a new accountant to advise us on tax matters, I reached out to Merrilee. I had watched her demonstrate her expertise for many years while serving on the board of directors alongside of her. Shortly thereafter, my partners and I hired Merrilee as our firm's CPA.

Serving on boards of directors is another proven technique for demonstrating our professional expertise. It is effective in a wide range of professions, from accounting to law to banking. Serving on boards is a great way to meet people. But it's more than that; it's a way for us to demonstrate our professional expertise to others.

There are many types of boards, and they are all looking for talented professionals. There are nonprofit boards, school boards, municipal boards – each seeking a broad range of professional expertise. There are also trade association boards, such as the American Association of CPAs and the American Institute of Architects.

Chuck McDonald is the South Carolina attorney we met earlier who specializes in serving the construction industry. Chuck is a member of the South Carolina Bar Construction Law Section, having served on its executive committee for a number of years and as chairperson of the section. He is also a member of the American Bar Association's Forum on the Construction Industry. Chuck's board roles serve as signals that he is a well-respected member of his professional community.

There are opportunities with for-profit companies as well – both privately held and public companies. These are not easy positions to land early in our careers, but opportunities can present themselves once we've established a professional reputation and built our professional network.

If serving on a board of directors appeals to you, I encourage you to get involved. Choose an organization whose mission resonates with your personal values. Like all of the other techniques we've discussed, serving on a board requires a time commitment. In some cases, the time commitment can be significant. But, as we learned from Merilee and Chuck, serving on boards can assist prospective clients in their decision-making journeys.

Technique 6: High-Profile Work and Case Studies

Clients prefer to work with professionals who have significant experience in solving problems similar to their own. In many cases, the best way to demonstrate our expertise is through our previous work for similar clients. Jean Brillat-Savarin, the nineteenth-century French politician, once quipped, "Let me taste the dish and you can spare me the rhetoric of how well you cook."

Consider these scenarios:

- If you were a community looking to design a new high school, wouldn't you seek the help of an architectural design firm with an extensive portfolio of school projects?
- If you were a private company about to enter Chapter 11 bankruptcy, wouldn't you seek the help of a corporate bankruptcy attorney with a long resume of similar client work?

- If you were a century-old consumer goods retailer in need of a branding overhaul, wouldn't you look for someone who had turned around other tired, struggling brands?

The work you have done for others becomes your reputation. Longfellow noted, "We judge ourselves by what we feel capable of doing, while others judge us by what we have already done." Many successful rainmakers leverage their high-profile work to open the doors to similar client work. These marquee projects become a calling card for future clients.

I realize that not every profession lends itself to this approach. There are certain types of services that are more discreet. If you are a communications specialist helping troubled companies navigate crisis PR situations, your clients may not be too keen on you broadcasting their troubled situation to the world. But these situations are the exception rather than the norm.

Our websites and other marketing materials are the perfect place to showcase our most impressive work. This isn't bragging in a way that Ms. Coby would frown upon. This is simply sharing the impressive work we've done for others. I'm surprised at how infrequently professionals showcase their best work. If you were a prospective client researching potential providers, would you rather visit a firm's website that:

- Option A: Offers a lot of puffery about how awesome they are?
- Option B: Offers examples of the awesome work they've done for others?

I don't think I'm alone in preferring Option B. Option A is simply bragging. Bragging is making grandiose claims without any credible evidence to support it. Not only is bragging distasteful, but it's not very effective.

According to my research data, two-thirds of all new client inquiries arrive via company websites. These are inquiries from individuals with whom you do not have an existing relationship. Prospective clients do a lot of research online before deciding to reach out to us.

Clients are becoming more like Willard Vandiver every day. Willard who? Willard Vandiver is the late congressman who gave Missouri its famous tagline, the *Show Me State*. In an 1899 speech he gave at a Navy banquet in Philadelphia, he said:

> I come from a state that raises corn and cotton and cockleburs and Democrats, and frothy eloquence neither convinces nor satisfies me. I am from Missouri. You have got to show me.

Individuals and firms who demonstrate their expertise through good examples of their work build credibility. So I would say to all aspiring rainmakers, use your websites and social media platforms to showcase your great work. Keeping your website content fresh with your latest work provides prospective clients with channel markers that guide them safely to a good decision.

Technique 7: Industry Awards

Megan Armstrong Wold is the founder and president of Armstrong Marketing Solutions (AMS). AMS is a strategic brand and marketing firm in the Pacific Northwest. AMS and my firm had neighboring suites in an office building for many years.

Megan assembled a talented team of strategists and creatives around her. And her efforts paid off. It wasn't long before AMS began winning advertising and industry awards. AMS recently announced that it had been recognized for work on a client's integrated brand identity campaign with a Silver ADDY Award at the American Advertising Awards Northwest competition. AMS was also recognized by the national Craft Beer Marketing Awards for their work with a large regional craft brewery.

Being recognized for your great work by an industry association is a huge boost to a firm's reputation – both by peers and clients. It's recognition by industry peers that you're doing excellent work. Clients often struggle to tell who the real experts are. Often there is a large gap between the client's expertise and your own as it relates to the work you do. When your work receives an industry award, it's a strong indication that the work you're doing is really good.

Many professions have awards for the work done by the members of their communities. If your industry is among them, I encourage you to investigate it more closely. The reputational boost sends a strong signal to the marketplace that your firm is worthy of serious consideration.

Technique 8: Professional Certifications

Mike McCracken is a jet aircraft broker in Tampa, Florida. His company, Hawkeye Aircraft, specializes in helping companies and individuals find and purchase jets for corporate and personal travel. Mike is a seasoned aviation professional with over 35 years in the aircraft industry. He has been recognized numerous times for his consistent success, including the Beechcraft Chairman's Award, Beechcraft Olive Ann Beech Award for King Airs, and Hawker Legion of Honor for Highest Total Units.

Clearly, Mike's in a niche business and most of us will never be in the market for a private jet. If we are, Mike's a guy you need to know. If you weren't convinced of Mike's qualifications based upon his impressive bio, maybe this would seal the deal: Mike is certified as an American Society of Appraisers Senior Aircraft Appraiser and is certified by the Appraisal Qualifications Board to do USPAP-compliant appraisals.

Most professions have professional certifications that can help signal to prospective clients that you are a true expert of your craft. Clearly, you will list all of your certifications and qualifications on your personal bio. But in order for a credential to be of significant value, it has to be unique. If everybody has it, it really doesn't do you much good in helping to differentiate you in the client's mind.

If you are a practicing accountant, you have your CPA (Certified Public Accountant) license. Without it, you wouldn't be able to practice legally, and the same goes for law, medicine, engineering, and architecture. Obviously, you will list these credentials on your webpage. But having your license doesn't set you apart from the pack.

If, however, a prospective client suspects that an internal employee is embezzling money from the firm's coffers, being a Certified Fraud Examiner will certainly get their attention. Professional certifications must tie into your special niche of work if it is to be of much value. The Certified Fraud Examiner is a handy certification to have if you specialize in helping clients catch crooks.

Depending upon your profession, certifications can be a highly effective way of demonstrating your expertise to prospective clients. Many of these certifications require a significant time and financial commitment. In addition to the up-front commitment, many certifications require annual education to stay current. Notwithstanding these hurdles, having a special certification can lead to more business for many of us. If it sets us apart from others, it could be the final nudge a prospective client needs to decide you are the real expert.

Choosing Where to Focus Your Energy

So which of these proven approaches will work best for you? John Zombro has nearly 30 years as a peak performance coach for athletes. John is a prolific writer in the field of human physical performance and host of *The Lifetime Athlete* podcast. When asked which techniques are best at demonstrating one's expertise, John replied:

> The method you put the most time into, are best at, and enjoy the most is going to get the most results.

When thinking about which ways you choose to demonstrate your professional expertise, remember the advice I offered our daughter when she entered high school: *You can do anything, but you can't do everything.* So my recommendation is this: It's better to pick one or two and be consistent than to dabble a little in everything. Your choices will be influenced by your profession and shaped by your personal skills and preferences. And, through honesty and consistency, you'll become a thoughtful guide in the client's buying decision journey.

References

Noel Sobelman story: Interview by *How to Win Client Business* research team, 2020.

Amir Tohid story: Interview by *How to Win Client Business* research team, 2020.

Billy Newsome story: Interview by *How to Win Client Business* research team, 2020.

Mike McCormick story: Interview by *How to Win Client Business* research team, 2020.

Merilee Glover story: Inspired by real characters known by the author.

Chuck McDonald story: Interview by *How to Win Client Business* research team, 2020.

"The 7 Most Common Client Pathways." Survey conducted by Fletcher & Company, LLC, 2020. https://www.fletcherandcompany.net/new-client-pathways-survey-findings/.

Willard Vandiver story: Phyllis Rossiter. "I'm From Missouri: You'll Have to Show Me," *Rural Missouri* 42, no. 3, March 1989.

Megan Armstrong Wold story: Interview by *How to Win Client Business* research team, 2020.

Mike McCracken story: Interview by *How to Win Client Business* research team, 2020.

11

Using LinkedIn to Build Your Credibility

It Won't Make the Cash Register Ring, So What's It Good For?

Being good at things is the only thing that earns you clout or connections.
—Steve Albini, record producer for Nirvana

Master Digital Marketing at Oxford

Just the other day this popped up in my morning newsfeed:

Marketing in the digital age requires a keen understanding of the media landscape – and a firm grasp on the fast-moving, technology-enabled marketplace. To succeed in such a dynamic environment, marketing professionals need to stay abreast of the latest trends and technologies – so they can develop winning strategies for their clients and organizations.

Clearly, I need to master digital marketing if I'm going to succeed in the twenty-first century, right? I want in on this – wait, here's my credit card.

Before we rush off to become certified as a digital marketer, let's heed the advice of ESPN's Lee Corso: "Not so fast." While digital marketing may be hot, trendy, and attention-grabbing, it's not meant for you and me – those of us working in professional services. It's meant for those selling sneakers, laptop computers, and ergonomic office chairs – things that people will actually buy on the internet. So don't be suckered by these teaser ads; you don't need to attend a course at Oxford or anywhere else to learn the latest in digital marketing.

As it relates to the professional world where you and I live, mastering digital marketing shouldn't be high on your to-do list, unless you want to sell your personal collection of cute puppy photos. What should we be doing, then, as it relates to the digital world? As it turns out, the internet can do many valuable things for our professional success, just not in the ways that we may at first think.

What Is the Internet Good For?

The internet is the bridge between *demonstrating your expertise* and *building your professional ecosystem*. When used effectively, it is incredibly helpful in sharing content that signals your go-to expertise, and in building connections with those you wish to serve.

At its most basic level, the internet provides two essential functions:

1. **Sharing information**
2. **Connecting people**

These two functions make e-commerce possible. The internet facilitates the connection between interested buyers and willing sellers, and allows for the sharing of important information. Crowd-sourced ratings (such as Yelp) allow customers to vet the reputation of sellers. The internet makes possible large marketplaces like Amazon, eBay, and Craigslist.

Web marketplaces are highly successful at selling products. With the click of a mouse, we'll quickly buy a book, a coffee table, or a pair of shoes, possibly even a new car. Intuitively, one would think that the internet would serve as a marketplace for professional services as well. If we'll buy a car online, wouldn't we do the same for the services of a tax accountant, financial planner, or marketing strategist? For the work that we do, I'm disappointed to report this won't happen with the click of a mouse.

Will the day ever come when clients click on a link to hire us? I'm guessing it won't happen in our lifetime. The reasons by now are becoming clearer to us. Professional services are purchased when a relationship exists between those who need help and those who provide the help. And, as wonderful as the internet is, it takes much more than a good photo and catchy marketing for a prospective client to feel confident in hiring us.

LinkedIn Is a Powerful Enabler, But It Won't Do the Heavy Lifting for Us

If clients need to come to know us, respect our professional expertise, and trust that we are honest and dependable, what is the best way for us to utilize the internet? While Facebook, Twitter, and Instagram have their purposes, they aren't that useful to most of us in our professional lives.

Facebook is great at staying in touch with friends and family, and can be helpful to those in retail. Twitter is great for connecting with fans – if you're JLo or Tom Brady. And Instagram is awesome at visually sharing the work of a Portland artisan leather crafter. For you and me, harnessing the internet's power means LinkedIn and our professional websites.

It just so happens that the same things that the internet does well – sharing information and connecting people – can also be very useful for us. The internet helps us share our expert knowledge with others, and allows us to connect with people we want to know. It won't do the heavy lifting for us, but it can lighten the burden.

The internet won't make you an expert trial attorney or a talented executive recruiter. Your website won't write papers for you or speak at conferences. It won't win you professional awards (unless you're a web designer). It's up to you to become really good at your craft. The internet is only a tool to help you share your work with others, and connect with those you wish to meet.

Here are a few examples:

- If you were recently chosen as board chair of the local American Marketing Association chapter, post this on LinkedIn and update your web profile.
- If you recently gave a TEDx talk on the future of energy-efficient home design, post a link to the video online.
- If your team recently won an industry award in *Advertising Age* for most innovative PR campaign, share this on the web.
- If you passed Level 3 to become a Chartered Financial Analyst, add this to LinkedIn and your web bio.
- If you are teaching accounting at your local university, share this on your LinkedIn page.
- If you're scheduled to be an upcoming guest on Wharton Business Radio, post the link online.

In sharing all of these accomplishments on LinkedIn and your website, you provide signals to your audience that you're a respected member of your profession.

Don't Turn into Human Spam

This brings to mind a question many of us have: "How often should I post things on LinkedIn?" And while it doesn't have one simple answer, I think Austin Kleon's advice is a good guide: Don't turn into human spam. Austin offered this perspective in his helpful book *Show Your Work: 10 Ways to Share Your Creativity and Get Discovered*:

> Human spam is everywhere, and they exist in every profession. They don't want to listen to your ideas; they want to tell you theirs. They don't want to go to shows, but they thrust flyers at you on the sidewalk and scream at you to come to theirs. At some point, they didn't get the memo that the world owes none of us anything.

Austin adds, "If you want to be accepted by a community, you have to first be a good citizen of that community. If you're only pointing at your own stuff, you're doing it wrong. If you want to get, you have to give. If you want to be noticed, you have to notice others. Be thoughtful. Be considerate. Don't turn into human spam."

I think Austin hits the right tone. We're engaging in a community. We're sharing our work, and we're celebrating the successes of others. Austin explains that our web presence falls on a spectrum from hoarder to spammer. In the middle is the sweet spot: contributor.

It feels right to me to share my work online about once every week or two. I do this through my blog and LinkedIn. I see others who are sharing daily and others who never share. You'll find the right balance if you aim to be a contributor and not a spammer. LinkedIn isn't Facebook; you're not sharing every little moment of your day. If you have something to offer, then share it. If you find something helpful that someone else has shared, then share it with others.

The Power of Sharing and Connecting

Because prospective clients struggle to tell who the true experts are, sharing our achievements, credentials, and awards helps them feel more confident in their decision. It helps them see the vast knowledge inside of our heads. And it's a peek into their future world – a glimpse at what it might be like to be a client of yours. Sharing our past success helps others avoid the feeling of anticipated regret.

Not sharing our work online puts us at a competitive disadvantage relative to those who do. If you feel you're bragging about yourself, shift your mindset. You're not bragging – you're providing a valuable service to those in the market for your services. And Ms. Coby would approve – as long as we're 100% honest.

LinkedIn is also a useful tool for connecting with others we wish to meet. This is a natural segue into our next rainmaker skill: *building your professional ecosystem*. In the upcoming chapters, we'll learn how the previous rainmaker skills of *creating your personal brand identify* and *demonstrating your professional expertise* help connect us with those we wish to serve.

The Rumor That Isn't True

I'll occasionally hear from an aspiring rainmaker something along the lines of: "One of our firm's most successful partners is Martina. She doesn't write or do public speaking. She doesn't teach a university class or host a weekly radio program. She serves on no boards and has no special certifications. She doesn't even have a LinkedIn page."

In my experience, people like Martina are rare – perhaps very rare, like the elusive Hochstetter's butterfly orchid thriving on a single tropical mountaintop on the Azorean island of São Jorge.

What I commonly find is that a junior partner doesn't always see the full picture of Martina's long career. While it may be true that Martina doesn't do any of these things today, what we may not see are the decades of commitment to these activities before she became established.

When you dig a little deeper, you often see that Martina served as board chair of the hospital when she was in her 40s or that she taught business law for 10 years at the local university. Or that she had a weekly column in the state's newspaper – these things have long been forgotten.

Occasionally, however rare, two conditions may exist that allows a rainmaker to succeed without committing to credibility markers:

- Condition A: Your target audience is local.
- Condition B: The market is not overly competitive for your type of work.

Your Target Audience Is Local

For word of mouth to work well, the professional network has to be tightly knit. People need to know one another well, mingle often, and share information with each another on a regular basis. In local communities, professions such as banking, accounting, financial planning, insurance, and law are intertwined like the deep roots of an old oak. These professions lean on one another to keep their local economies operating smoothly. While this can exist on a national or global basis, it is the natural fabric of local communities.

The Market Is Not Overly Competitive for Your Type of Work

If the market for your profession is not highly competitive, it may be possible for you to succeed without committing much effort to demonstrating your expertise, in much the same way that it's much easier to sell BMWs if you're the only dealer within 500 miles. However, it's rare that you're the only service provider capable of doing the work. More often than not you'll be looking for ways to tip the balance in your favor. And even if your market isn't overly competitive today, there's no guarantee that it will remain that way.

While rainmakers may exist who don't actively demonstrate their expertise through credibility markers, they are far from common. Much more likely are the professionals we observe publishing articles in trade magazines, speaking at industry meetings, teaching classes at local universities, and serving on professional boards. Not only are these techniques highly effective at building our reputations, but they also bring meaning to our lives as contributing members of society.

So to the aspiring rainmaker out there who asks, "Martina doesn't do it, so why should I?" I would say that the odds are stacked in your favor when you do. From my experience, it's time well spent. When we provide those we wish to serve with strong examples of our expertise, we enhance our professional reputations and increase our chances of success.

References

Master Digital Marketing at Oxford: Inspired by real email received by the author.

Don't Turn into Human Spam: Austin Kleon. *Show Your Work! 10 Ways to Share Your Creativity and Get Discovered*. Workman Publishing, 2014.

The Rumor That Isn't True: Inspired by real characters known by the author.

John Roach. "Extremely Rare Orchid 'Rediscovered' on a Remote Island." NBC News, December 10, 2013.

Skill 3: Build Your Professional Ecosystem

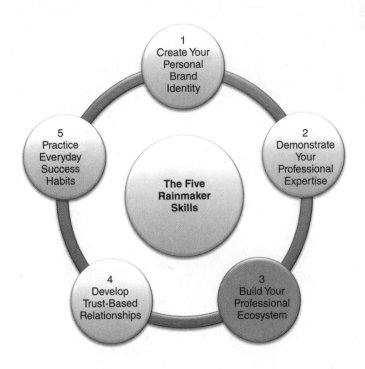

CHAPTER

12

The Two Hundred People You Need to Know

The Closest Thing to Knowing Something Is Knowing Where to Find It

> Build a good name. Keep your name clean. Don't make compromises. Don't worry about making a bunch of money or being successful. Be concerned with doing good work...and if you can build a good name, eventually that name will be its own currency.
> —William Burroughs, American writer and artist

Respect and trust are the currency of our trades. You may be a strategist, litigator, auditor, or designer by profession, but at the end of the day we're all in the relationship business. And the foundation of authentic relationships is respect and trust between those who need help and those who can provide help.

Each of us has roughly 200 people who would make all the difference if we knew them. Not 2000 or 2 million – but 200. Having a close relationship with these 200 people could make our careers.

For Dominic Barton – the former managing director of McKinsey & Co. we met earlier – the number is 500. When Dominic reads, he imagines that he is reading for his 500 friends and colleagues. If he finds something that could be helpful to one of them, he sends a handwritten note along with the article or book.

Whether your number is 200 or 500, our professional world is much smaller than we might at first realize. Our relatively small network of professionals provides an enormous source of value for everyone involved – through the currency of shared respect and trust.

Genuine Relationships, Not LinkedIn Connections

Here's a simple exercise. Take a few minutes and scroll through your LinkedIn network. Put each person into one of two groups:

- Group One: Those you know well enough to ask a favor
- Group Two: Those you barely know at all

I'm willing to bet that at least 90% of those in your LinkedIn are people you barely know. My guess is these are people who sent you a LinkedIn request and in less than two seconds you clicked "Accept." This group does not signify a meaningful relationship. Group Two is not your 200 people – at least not yet. Our 200 people are those we can call when we need help and those we assist when they need help.

The reason I make this distinction between Group One and Group Two is to make a point about relationship building. Our vast LinkedIn network doesn't equate to real-world relationships. Close relationships are built over time in getting to know someone, working side by side with them, and in giving and receiving assistance. Our goal as rainmakers is to build genuine relationships, not amass a gigabyte of names in an online Rolodex.

What's the Difference between Your Ecosystem and Network?

In much the same way that *selling* has become as welcome as a plaid coat at the used car lot, *networking* has become an unpleasant reminder of the card-swapping meet and greets at business happy hours. My palms are sweating now just thinking of the lapel tag in my desk drawer: "Hello! My Name is: DOUG."

We must differentiate between what we think of as networking and what our real aim is: building authentic relationships. Our goal is not to squirrel away 100 business cards in our coat pocket, but get to know people in a meaningful way.

OK, maybe networking doesn't make you uncomfortable. And if you enjoy business meet and greets, this is not meant in any way to diminish your enthusiasm. You may have a gift in meeting new people that many of us don't. But I'm willing to wager that those among us who enjoy networking are as rare as the guy who loves cold-calling – or as elusive as the Hochstetter's butterfly orchid.

I've asked you to rethink a couple of preconceived mindsets since embarking on this journey together:

Mindset Shift One: Seeing ourselves as problem-solvers instead of salespeople

Mindset Shift Two: Viewing our professional accomplishments as credibility markers rather than bragging

To this I'll add:

Mindset Shift Three: Thinking of our professional network as authentic relationships rather than names in a database

This is why this rainmaker skill is called building your professional *ecosystem* rather than building your *network*. Calling it an ecosystem helps us remember what the goal of networking was originally intended to be: building tight-knit groups of people that can help one another accomplish great things. Along the way, the word became synonymous with budget hotel ballrooms, lapel stickers, and sweaty handshakes.

None of us are islands. Even if we were the next Marie Curie or Albert Einstein, we need others to succeed – and they need us in order to succeed as well. The world operates best when an ecosystem works together to solve important world problems and satisfy vital client needs.

We're simply one node in a network of mutually supportive professionals. And in order for us to succeed as rainmakers, we have to invest time in getting to know others in our professional community. These relationships connect us to a world of opportunity that allows us to perform our life's work.

Why Your Ecosystem Matters: Revisiting the Seven Client Pathways

The importance of your 200-person network becomes clearer when we overlay this with the client pathways. You may recall the seven client pathways from Chapter 3, ranked from highest to lowest success rate:

1. Repeat business (from a satisfied client)
2. Referrals (from a satisfied client, trusted colleague, friend, or acquaintance)

3. Inquiries (from someone you know)
4. Inquiries (from someone you don't know)
5. Warm prospecting (with someone you know)
6. Warm prospecting (with an introduction)
7. Cold prospecting (with no introduction)

Our current and former clients are important parts of our professional eco-system. If we've done great work for a client in the past, there is a much greater chance that they'll hire us again in the future or recommend us to others. They are the source of the majority of the work that we do, and the number one source of referrals to new clients.

According to my firm's research, referrals and inquiries from our professional ecosystem represent over half of our new client business. Add to this warm prospecting with those in your ecosystem and we're up to two-thirds. Include warm prospecting with an introduction from someone in your ecosystem, and we're at over three-quarters of all new client business. This is worth repeating:

75% of all new client business originates from your professional ecosystem.

The most successful rainmakers invest significant time in building their professional ecosystem. They build their ecosystem and nourish it in order to open up the most significant client pathways. They invest the time required to build authentic relationships, not just clicking "Accept" to a LinkedIn invitation.

The Makeup of Your Professional Ecosystem

The term ecosystem was first coined by British ecologist Sir Arthur Tansley in the 1930s. Tansley was a professor of botany at Cambridge and Oxford, and helped pioneer the science of ecology. His major contribution was seeing the connections between the individual species in a biological environment.

He saw how the interconnections of the natural world, when taken as a whole, created a system. For the first time, scientists began to study not only individual species, but how the individual species worked together to assist in the biological health of the whole system.

This biological concept provides a valuable construct for us to study our professional communities as well. As individuals, we make up one small part of an ecosystem that is interconnected into a larger, more complex economic web. In much the same way that biological species can have mutually beneficial relationships, we can thrive by helping others succeed and others thrive in helping us.

Professional ecosystems exist in every industry. A CPA helps her client by referring a reputable financial planner, and a tax attorney assists his client by providing a connection to an experienced accounting team. The same holds true when an architect introduces a client to an interior designer, and when a civil engineer connects an urban planner to the municipal director at city hall.

Or a Seattle web designer builds a website pro bono for a local nonprofit, and the nonprofit's board chair introduces her to his Fortune 500 employer. A commercial lender in Chicago introduces his manufacturing client to a customs broker in China, and the manufacturing client introduces the broker to an import specialist in Denver.

These scenarios happen every day in the world of commerce because the world operates smoother and faster with the currency of respect and trust. In the absence of mutual support, the world would plod along like a plow mule in a horse race.

So who should be included in our professional ecosystems? It depends. Many factors influence your most important 200 people: your industry, profession, location, and target audience. Consider these potential ecosystem partners as a starting point:

- Clients, current and former
- Colleagues in your company or firm
- Business partners/alliances
- Colleagues in related professions
- Industry association members
- Professional organization members
- Community officials
- Former classmates
- Friends and neighbors

While the list is almost limitless, the key is to understand how your specific profession works. Each profession has an ecosystem that thrives when individuals work together to support one another. While it is possible to be a lone wolf – working alone without the support of the pack – the most successful rainmakers work as helpful members of a larger community.

The Double Helix of Your Ecosystem and Trust

It's hard to have a discussion about your professional ecosystem without including the topic of trust-based relationships. Your ecosystem is intertwined with trust like the strands of the DNA double helix. For the purposes of our journey together, I've broken these into two separate sections. In Chapters 13 through 16 we'll take a close look at how to build your professional ecosystem. In Chapters 17 through 20 we'll closely examine the nature of trust-based relationships. Throughout the entire discussion, we'll see how trust and your ecosystem engage with each other in a continuous dance.

References

Dominic Barton story: Interview by *How Clients Buy* research team, 2017.

"Top Client Pathways." Survey conducted by Fletcher & Company, LLC, 2020. https://www.fletcherandcompany.net/new-client-pathways-survey-findings/.

Sir Arthur Tansley story: Sir Harry Godwin. "Arthur George Tansley. 1871–1955." *Biographical Memoirs of Fellows of the Royal Society*. Royal Publishing Society, 1957.

13

Does Cold-Calling Work? And What to Do if It Doesn't

Remember What Mom Said: *Don't Talk to Strangers!*

Your ability to build a rich, robust network of relationships with people who trust you is the barometer that predicts your success. It's as easy as that. It's also as hard as that, because relationship-building concepts like generosity, intimacy, accountability and candor are all easier said than done.

—Keith Ferrazzi, author, *Who's Got Your Back*

Mel vented, "No one wants to talk to me! When I call a prospect I get hung up on. Or the admin puts me into voicemail – of course, no one ever calls back."

"Why do you think that is?" I asked.

"I don't know. Maybe they're busy. Honestly, it drives me crazy. I feel like I'm wasting my time. And, frankly, I don't think I can do it any longer," Mel said.

I responded, "Do you ever take a call from a stranger?"

"I guess not. If I don't recognize the caller, I'll let it go to voicemail," Mel replied.

"Huh," I said, "I guess your prospects are not that different than you or me."

Cold-calling has the lowest success rate in winning new client business – representing roughly 12% according to my survey data. And if we looked at the return on investment of the time required to achieve that 12%, I'd say it's hard-won business. Just like my friend Mel, I have learned the hard way that cold-calling is difficult. And I haven't found it to be very effective in my practice, and I don't enjoy it very much.

If you're one of the rare individuals who have found success with cold-calling – and if it brings you a solid pipeline of business – then please don't let me stop you. I have heard of sales training programs that specialize in teaching people to be master cold callers. I'm sure these are useful programs, and I could certainly get much better at cold-calling. It's just that I don't want to; I believe there is a much better way for most of us, although there are a few isolated cases where cold prospecting may be necessary (more about this later).

I'm willing to bet that you don't like cold-calling people, either. And, aside from it not being a very enjoyable way to spend the day, it's a difficult way of finding new clients. Most of us would be better off searching the sands of India for the next Hope Diamond. Calling strangers hasn't proven to be a great use of our time for the vast majority of service professionals. Pattern-recognizing creatures that we are, we stop doing what isn't working. The trouble is this: If cold-calling isn't the best use of our time, what should we be doing?

Remember What Mom Said: *Don't Talk to Strangers*

Perhaps another reason why people won't take your call is that mom's voice still echoes in our heads: "Don't talk to strangers!" Maybe deep in our subconscious brain somewhere are the lessons we learned as kids. If so, then maybe there's hope for my own kids.

People generally don't gladly pick up the phone and chat away with a total stranger. Consider this scenario:

Dick: "Hi, you don't know me from Adam, but I'm Richard Pendarvis and I specialize in helping turn around struggling companies. You can call me Dick."

Prospect: "Hi, Dick. Great to hear from you; you called at the perfect time. I was just going over my profit and loss statement and, boy, are we ever struggling."

Dick: "Great! I mean, I'm sorry to hear that. Could I come over to your office tomorrow and share with you our approach to turning around struggling companies."

Prospect: "That would be awesome. How about 8:00 a.m.?"

Dick: "Perfect. See you then."

Have you ever had a cold call go like this? I'm guessing not. It hasn't for me, at least not often enough to justify sitting behind a desk dialing for dollars all day.

Great, thanks. You've told me what I already know. Cold-calling stinks. What next? Well, the answer is relatively simple to understand, but it will take a commitment of your time. But I think it's a far better use of your life energy.

Sticking with my childhood theme here, I'll share with you the second part of my Don't Talk to Strangers idea; it's called the Make Friends on the Playground concept.

Making Friends on the Playground

Have you ever noticed kids playing on the playground? OK, probably not, unless you have young kids of your own. Kids have no trouble making friends on the playground. You'll see a few kids playing together on the swings, taking turns pushing one another, seeing who can go the highest.

And you'll also see a few kids gathered for a kickball game. A few others will be playing tag. There may be a couple of kids chasing a butterfly. Perhaps a few stay inside and read or draw. You get the picture. Kids naturally make friends by gravitating to the things that interest them. I bet if you were a grade-school teacher you could predict before recess break which ones would be doing each activity. Kids have their favorite playground pursuits and the same kids gather together most every day.

As kids get older their interests change, but the themes remain. Kickball becomes soccer. The monkey bars become the climbing gym. Tag becomes track and field. Butterfly chasing becomes science club. Crayons become paintbrushes. Think back on your best childhood friends – how did you meet them?

Take this parable into our adult lives. As adults, we are just grown-up versions of our childhood selves. Our waistlines may have grown a bit and our hair has grayed or thinned, but, really, deep inside we're just older kids. And like kids make friends organically on the playground, we make friends in much the same way as adults.

If you're into gardening, you easily make friends with others in the garden club. If you're into golf or tennis, you have a list of friends to call when you want to play. Whatever your interests – cycling, computers, cars, cooking – you have a group of friends who enjoy these same activities. Making friends in this way seems effortless, because it is.

Would you pick up the phone and randomly call someone's house to say, "Hi, you don't know me, but I love trail running. Would you like to run with me tomorrow morning at 8:00 a.m.?" The response is likely to be "No, I don't know you and wouldn't like to run with you." So why do we try this approach to finding our clients? It really doesn't make a lot of sense. We'd be much better off finding clients in the same way that we found friends as kids. Yes, you may be able to win client business through cold-calling if you have the stomach for it. And I bet we could chop down a giant Redwood with a screwdriver.

Making Friends with Your Two Hundred People

In observing how successful people build their networks, I've witnessed how they make friends in a natural way, as we did as kids. For me, I'm 100% confident I'd rather build my professional ecosystem in this way than in calling total strangers.

If you buy into this approach, then what we need to do becomes clearer: you need to follow your professional interests – in much the same way a kid would gravitate to swings or butterflies – and you'll find your people. Or in Seth Godin's words, "you'll find your tribe." In *Tribes*, Seth offers:

> A tribe is any group of people, large or small, who are connected to one another, a leader, and an idea. For millions of years, humans have joined tribes, be they religious, ethnic, political, or even musical (think of the Deadheads). It's our nature.

Another wonderful thing about making friends in a more natural way is that it works for introverts and extroverts alike. If you're more introverted – like many of us are – I bet you'll find making friends in this way more suitable to your personality. Even the shyest of the shy light up when they find a kindred spirit, like Stephen Wozniak at a 1970s Palo Alto computer club gathering.

So I say to you: Go find your tribe. They'll be easy to find because they'll have similar interests as you. The good news is you have plenty of choices to pick from. Here are a few examples to consider:

- Trade associations and events
- Nonprofit organizations
- Charity events
- Industry conferences
- Special interest clubs
- Sporting events
- Weddings and birthdays
- Alumni reunions
- Community gatherings
- Neighborhood socials
- Peer-to-peer groups
- Seminars
- Continuing education
- Ad hoc groups
- Can't find one that interests you? Create your own group.

The amazing thing is that when meeting people in these types of settings, we're receptive to meeting other people. There's a common connection that serves as a natural ice breaker. The conversations will last much longer because we're actually enjoying ourselves. Often, it isn't until the conversation is nearing its end that we learn what the other person does. Here's a story about Mia and Bill from a local fund raiser for Engineers Without Borders.

Bill:	"It's been nice meeting you, Mia. By the way, what do you do for a living?"
Mia:	"I'm a computer engineer. I work for Cloud9 Consulting. We help small to medium-sized companies migrate their computers systems to the cloud – you know, the internet."
Bill:	"That's funny. My friend, Josh, was just telling me his company was thinking about doing that. I should connect you two."
Mia:	"That would be awesome. Here's my card. Oh, wait; I don't have one with me. Can I call you next week and give you my contact information?"
Bill:	"Please do. I think Josh would like to talk with you."
Mia:	"Thanks, Bill. I really appreciate the introduction. Talk to you next week."
Bill:	"Great, I look forward to it."

The following day Bill checks out Mia's LinkedIn profile. She has a BS in computer engineering from the University of Illinois. She spent 10 years working with Accenture and Microsoft. She is chapter president of Engineers Without Borders and teaches an online seminar in cloud-based computing at Seattle University. Bill is genuinely impressed by Mia's credentials. The following week Bill introduces Mia and Josh, and the last he heard she was migrating Josh's old servers to the web.

Let's look at what just occurred here, and you'll begin to see how the pieces of the rainmaker puzzle fit nicely together. First off, you could tell that Mia had a clear *personal brand identity*: She was a computer engineer helping small to medium-sized companies migrate to the cloud. She did a great job of providing credibility markers to those who may be interested in her services – *she demonstrated her expertise* through her education, work experience, and professional activities.

Finally, she was *building her professional ecosystem* in a natural way through an organization that interested her: Engineers Without Borders. She didn't spend her time cold-calling total strangers. When she met Josh – through the introduction from Bill – the conversation seemed effortless. It felt more natural and it was far more effective at growing her practice.

The one caveat to this is that our interest has to be genuine. We have to be sincerely interested in the activity or organization for it to be effective. When we're interested in simply padding our resume or trading business cards with movers and shakers, it won't work. Others will see us coming from a mile away. The instant you misrepresent yourself or your intentions is the instant you lose credibility. The benefits of making friends in a natural way only happen when our hearts are in the right place.

When Cold-Calling Works Best

I think you know where I stand on cold-calling. And I'm guessing you don't feel differently. I think it's much more effective to invest the time in building our professional ecosystem more organically. That said, there are some professionals who do practice cold-calling. And some have found success with it. I've studied this carefully and identified a few situations where cold-calling can work. It may not be the only effective approach, but in certain cases it may be among the best choices.

Services That Are Packaged as a Product

Cold-calling or prospecting can be effective when used with services that have been "productized." If your service lends itself to packaging into a product, then cold prospecting can be effective, in much the same way that salespeople sell products every day – everything from pharmaceuticals to network servers to medical equipment. If you do your homework first and you work at it every day, this approach can be successful.

Eric Gregg is the CEO of ClearlyRated we met previously. ClearlyRated is a software-as-a-service business – they provide a client satisfaction measurement program for accounting, legal, HR, and staffing firms. Yet its service fits well in the productized category using software technology and a proprietary process.

Eric shared with me recently that his team has had good success with cold prospecting. But he went on to say, "You have to commit to adding value to the person answering your call and stay patient and persistent. Our outbound sales efforts wouldn't work if we didn't have strong brand awareness and a willingness to share valuable industry specific research with our prospects. Cold-calling is most effective when you are creating value for the buyer from the very start."

When Opportunities Arise to Help Those Whom You Don't Know

Another situation where I have seen cold-calling work effectively is when you have a unique opportunity in mind for a specific company or individual. This scenario is common in the real estate, finance, and other transactional services.

Graham Anthony is the founder of Anthony Advisors, a mid-market mergers & acquisition (M&A) services firm. Over the last 20 years, Graham has facilitated the sale or acquisition of roughly 50 mid-market companies – typically in the range of $10M to $100M in revenue. As a general rule, cold-calling is not something that Graham practices very often. He shared that the majority of his business comes from repeat business and referrals. He did, however, describe a situation where he does reach out to someone cold.

Graham has used cold-calling when facilitating the sale of a company. He said that in some cases, there are a few strategic buyers that logically make sense for a deal – perhaps a larger competitor that is in the same market, or a company in a different geographic region that may be interested in a new territory. In situations like these, Graham will contact the strategic buyer to explain the acquisition opportunity. He said that cold-calling has proven effective at putting together successful M&A transactions.

Graham offered this important perspective: "The key part is understanding the wants and needs of the person I am calling and helping solve their needs. If one can show a buyer or a seller how a proposed transaction moves them closer to what they want and away from what they don't, they will strongly move toward the opportunity. The focus here is what is in it for them. If one understands their hopes, fears, and dreams, one can help provide them with a solution. One has to establish and maintain trust through the process for the above approach to work."

Your Firm Solves a Problem or Provides a Solution That Prospects Aren't Aware Of

For some specialty services, the prospective client is unaware of their problem. Or, alternatively, they are unaware that your service even exists. It's a big, wide world out there, and among the tens of thousands of companies in the U.S. some solve problems that are new or undiscovered by their target audiences.

One example that comes to mind is Zquared, a marketing services firm that helps companies sell their products on Amazon. It's a relatively new business – less than 10 years old – and they have figured out how to optimize the Amazon sales channel. Zquared helps its clients manage all aspects of the Amazon platform: design, fulfillment, product listings, enhanced brand content, headline and sponsored ads, and social media marketing. If you're a company that sells products on Amazon, you need to know about Zquared's services. The problem is that you often don't, because it's a relatively new path that isn't well worn.

In a recent conversation I had with Keith Latson, Zquared's business development director, he explained that in the early going his team did a lot of cold-calling – and found success. "The key," Keith explained, "is in researching prospects so well that you know as much about the company's products and markets as they do." He continued, "we were trying to demonstrate that we had done our homework and built a strong case for deserving a seat at the table."

Keith's team had a specific client profile in mind. They searched the internet to find the companies that fit their target audience. The second piece of advice that Keith offered was patience. He said the client's buying decision journey was often up to a year or longer – and you must let the relationship develop gradually and build momentum as trust was established.

When a prospective client is unaware that a service exists, they are unlikely to ask a trusted colleague for a referral, or even think to inquire about the service. It's so niche that the referral and inquiry pathways don't easily flow like they may in other professions. In situations like these, cold prospecting may be one of your only choices. Such is the world that we live in. If business were easy, everybody would own a Rolex and drive a BMW.

Cold-Calling and Making Friends

We've examined several examples of when cold-calling may be an effective approach to new client business. For those individuals and firms who have found success with this pathway, they quickly offer that it is often a long sales cycle. As we've learned, there are many milestones in the client's buying decision journey, and with cold prospecting you are literally starting from square one. Additionally, it helps to have a very specific target market and a compelling story to tell in order to break through the noise.

For the vast majority of us, cold-calling isn't the most effective client development approach. We'd be much better off remembering how to make friends the way we once did as kids. Somewhere along the way we forgot how. We graduated from college, bought a suit, and evolved into some alien creature like the Klingons in *Star Trek*. If we can just unlearn some unnatural behavior and relearn how to make friends again, we'll find much more success and enjoyment in our professional lives. Unlearning things is sometimes harder and more important than learning new things.

References

Mel story: Inspired by real characters known by the author. Names and locations have been changed.

"Top Client Pathways." Survey conducted by Fletcher & Company, LLC, 2020. https://www.fletcherandcompany.net/new-client-pathways-survey-findings/.

Seth Godin. *Tribes: We Need You to Lead Us*. Penguin, 2008.

Bill and Mia story: Inspired by real characters known by the author. Names and locations have been changed.

Eric Gregg story: Interview by *How to Win Client Business* research team, 2020.

Graham Antony story: Interview by *How to Win Client Business* research team, 2020.

Keith Latson story: Interview by *How to Win Client Business* research team, 2020.

CHAPTER

14

Making Friends
in a Natural Way

How to Get an Introduction without Seeming Pushy

For Dad, life was good, and what's more, it was always just about to get better. He awoke happy, and every day dawned on a wealth of opportunities that beckoned his ever-curious mind to visit new places, to meet new people, to discover new adventures.
—Jim Nantz, American sportscaster and author of *Always by My Side*

It's not far-fetched to say that David Maister has influenced our professions more than any other person. David was the first academic heavyweight to focus attention on accountants, lawyers, consultants, and others like us.

Until David came along in the 1980s, business scholars studied General Motors, Johnson & Johnson, or similar companies making and selling things such as cars, medicine, and household goods. Clearly important work, but academics overlooked people working in offices rather than manufacturing plants. David's specialty was studying companies whose assets walked in and out of buildings each day – companies that manufactured ideas, not widgets.

When David left his teaching career at Harvard Business School, he was relatively unknown, except perhaps in academic circles. He went on to write many best-selling books that line the shelves of glass-and-steel office towers throughout the world; *The Trusted Advisor* is the most well-known. My well-thumbed copies of his books are within arm's reach every day.

Now, the plot thickens. When I was in my first year of graduate school, I sat next to Cliff Farrah. We were arranged alphabetically in our amphitheater-style classroom, and Fletcher came right after Farrah on our class roster. For the first semester, Cliff and I sat side by side every day. At Darden, the students stay in place throughout the day while the teachers move classrooms, which actually made things a lot easier. Rather than 240 students scrambling down the halls every 90 minutes, you had a few dozen professors moving between classrooms.

As you might expect, you get to know someone plenty well when you spend thirty hours a week together. On the surface, you wouldn't think Cliff and I had much in common – a city kid from Boston, and a small-town boy from South Carolina. But we soon discovered that we both shared a love of sailing and the sea. And somewhere in the fertile mix of shared late nights, exam prep, job interviews, and weekend beers, we became good friends.

If the story ended here, it wouldn't be worth sharing. But it gets more interesting. Cliff worked for David Maister in Boston before grad school. Cliff was David's right-hand man assisting him with everything from research to client engagements. I knew nothing of David's work before grad school. Through Cliff, I became well acquainted with David Maister and the field of professional services.

Fast forward six years and I find myself in Kittery Point, Maine, attending Cliff's wedding. At the wedding reception, it just so happened that I was seated next to David Maister. Well, I'm sure it didn't happen by accident – Cliff had a hunch the two of us might have something in common. Frankly, I was star struck in meeting David, the way a Harry Potter fan might feel meeting J. K. Rowling.

I had recently co-founded North Star Consulting Group, a marketing research firm specializing in large-scale business surveys. David was working at the time with Omnicom, the New York-based global advertising services firm. Over dinner and wine, David shared that he was looking for a marketing research firm to assist him with a global employee survey at Omnicom. You see where this is leading, right?

Without dragging this story out too much longer, I'll quickly tie up the loose ends. Yes, my firm went on to assist David with the Omnicom project. We helped David with the research for the book that came out of this study, *Practice What You Preach*, and North Star continues to serve Omnicom to this day.

How did a boutique firm end up working with the largest advertising firm in the world? Well, in retrospect, it's easy to see through the lens of the client's decision-making journey. Clients work with people with whom they have relationships, or in many cases a chain of relationships. My relationship with Cliff led to my relationship with David, which led to North Star serving Omnicom. This chain of relationships served as a means of establishing mutual respect and trust.

Would we have landed this opportunity if I had simply called up the chairman of Omnicom? "Hey, Mr. Chairman, you don't know me, but I'm Doug Fletcher, co-founder of a marketing research firm." If I were I betting man, I wouldn't put my money on it.

Ok, great. So you're telling me I need to go to grad school and attend a lot of weddings? No, the moral of the story is that there is power in the professional communities we build. Rainmakers develop webs of talented people that help one another succeed. Just as we learned from Sir Arthur Tansley, an ecosystem is stronger when each species thrives, and each species thrives in helping one another. This story also illustrates the way in which our ecosystems sometimes connect in random and absurdly serendipitous ways – more often than one might think.

Warm Prospecting: Leveraging Your Relationships to Make New Friends

The best way to build our ecosystem is through activities that interest us: boards, associations, clubs, nonprofits, continuing education, and the like. In this way we'll make friends organically and build authentic relationships. With a clear personal brand identity and strong credibility markers, we'll set ourselves up for meeting the people who can help influence our professional careers.

We've established that cold-calling isn't much fun, nor is it highly productive. But there is a twist to cold-calling that does work and feels natural to us: the practice of warm prospecting.

Warm prospecting is simply making new friends through introductions from your existing relationships. While cold-calling feels awkward, warm prospecting feels more natural, in much the same way that we made friends as kids.

Thinking back to childhood, we often made new friends through our existing friends. Imagine this scenario: while hanging out at your best friend's house, another kid stops by. The conversation may have gone something like this:

New Kid:	"Hey. What're ya doing?"
Your Best Friend:	"Not much – just goofing around."
New Kid:	"We were thinking of playing soccer at my house; wanna play?"
Your Best Friend:	"Awesome. What time?"
New Kid:	"Oh, I don't know, maybe after lunch?"
Your Best Friend:	"Cool. I'll be there. Do you mind if I bring my friend, Jim? He loves soccer, too."
New Kid:	"Sure! We need enough kids for two teams."
Your Best Friend:	"Thanks. We'll be there!"

Whatever you and your childhood friends were into, that's the way we made new friends. Through our best friend we now have a new friend, who happens to have similar interests. Next time I want to round up a game of soccer, I'll call up the new friend and invite her, too. It's not a weird or uncomfortable conversation at all.

As adults, we have a lot to learn from watching kids make new friends. We can make new friends in similar ways. When we want to meet a new person, we often have an instant connection through a mutual friend with similar interests. This feels far less awkward than calling someone up out of the blue. And it's a lot more effective; our survey data suggests this approach is roughly twice as successful as cold-calling.

When we build close relationships with our 200 people, this opens up a whole universe of other people we get to know. Because every one of our friends or colleagues also has a network of close relationships, you are now one relationship away from a thousand or more people. While calling up a total stranger is unpleasant and not very effective, calling up a friend of a friend isn't so bad. especially when we have a personal introduction from our mutual friend.

Some of us may find it uncomfortable to leverage our friendships in meeting new people. You shouldn't – good friends help one another, as long as you're willing to do the same for them. This may require a mindset shift as it relates to building our ecosystem:

Mindset Shift Four: Using our professional ecosystem to make new friends is highly appropriate when it's based upon a genuine desire to help others.

You can see why having a clear personal brand identity and strong professional credibility is essential to building your ecosystem. In order for others to be in a position to help you, you have to have clear expertise and have a sincere desire to help others succeed.

Consider these two examples:

1. Hi, Fran. This is Joe. He's an architect and he'd like to meet you.
2. Hi, Fran. I'd like to introduce you to my friend Joe. He specializes in energy-efficient home design. We serve together on the regional certification board. You may have seen his recent article in *Architectural Digest*. He's interested in having coffee with you to learn more about your work.

Which of these scenarios is more likely to engender a positive reply? In order for others to help you, you first have to be worthy of an introduction. It starts with rainmaker skills 1 and 2.

Nobody Likes Human Spam

Remember the warning from Austin Kleon in Chapter 11: *Don't turn into human spam*, wanting everyone to help you when you're unwilling to help others. This is especially true when building our professional ecosystem. In order for others to help us, we have to be helpful to them. It's like magical dust that is sprinkled upon our professional relationships. Over time, those that come to know you can sense when you're out for number one. Ralph Waldo Emerson may have said it best: "Your actions speak so loudly, I cannot hear what you are saying."

Using our ecosystem to build new friendships works well when our friends respect our work and trust that we have a genuine desire to help others. And when we make new friends through our personal community, it's highly effective – as simple as making new friends as kids.

Let me give you an example of human spam. Actually, it's real spam. Not the canned Spam meat, but the email kind we abhor. Consider this email I received today. I'm sure you get about a hundred of these each day as well. The name and company have been changed to avoid calls from an attorney.

Subject: Doug, phone call on Friday at 1:00 p.m.?

Hi Doug,

My team found your company online and I think our approach to finding and generating new clients might work for you. Our solution combines calls, emails, LinkedIn, and social media to make targeted and timely touchpoints.

We'd like to confirm via a short consultative meeting, which will give us enough information to come up with a proposal for a concrete marketing plan.

Can we go for a phone call at 1:00 p.m. your time? Please let me know if that works for you. Otherwise, please suggest a better day and time that fits your schedule.

Best,

John Doe

B2B Coordinator | Horizontal Networks

This email is just as fruitless as cold-calling. It may cost less, but the effect is equally annoying and ineffective. I don't know John Doe or Horizontal Networks. I don't have any reason to respect or trust John or his company. I'm not going to invest the time to research his credibility online. And I'm certainly not scheduling a call with him on Friday at 1:00 p.m.

Another example is social media spamming. Have you ever clicked "accept" to someone's LinkedIn request to quickly find them pestering you with offers? You don't know them, you don't respect them, and you certainly don't trust them, and they're not interested in building a relationship. They want you to buy something from them – today!

We heard from Judy Selby earlier – she's the attorney in New York who writes frequently as a means of building relationships. Judy recently shared this with me:

"It is so alienating when someone connects with you on LinkedIn and then starts trying to sell you something right away. It's such a turn-off."

Guess what? Our prospective clients are pretty much like you and me. We don't like cold calls and we don't like email or social media spam – and neither do our prospective clients.

How to Get an Introduction without Seeming Pushy

It's wonderful when someone recommends or introduces us to a prospective client. Genuine introductions and referrals carry respect and trust like electrons through a copper wire. It assists us in building our professional ecosystems effortlessly and organically.

These types of introductions don't always happen on their own. What if we want to meet someone who shares a mutual connection with a friend or colleague? I'll share with you my approach. My approach may not be the only way, but it's the only approach that feels comfortable to me.

I never ask for a recommendation from a colleague. If recommendations are to work effectively, they have to be heartfelt and given freely. To me, it's like kindness. You can't make someone offer you kindness, and it's not very effective to ask someone to be kind to you.

Many of the most successful professionals I speak with share a similar philosophy. One example is Charles Moren, a senior sales engineer with R.L. Matus & Associates headquartered in Charlotte, North Carolina. Charles recently revealed that he "never asks for recommendations or referrals."

If you help solve someone's problems, they will come back again the next time they have a similar problem. And if you help them often enough, you will begin to build a personal reputation for what you do. Word will spread naturally among the people in the industry – and that is the best way.

If someone is a raving fan of yours – and we all have at least a few – then recommendations will happen organically. Following the rainmaker skills, we'll receive more than our fair share of referrals when we:

- Create a strong personal brand identity
- Demonstrate our professional expertise
- Build our professional ecosystem
- Develop trust-based relationships

For me, that's as far as I'll go in opening up the referral or recommendation pathway. Like kindness, referrals work best when they are given without asking.

I realize there may be some among us who readily ask for a referral. I feel more comfortable in providing well-deserved referrals to others without expecting anything in return. As with kindness, I have experienced that it comes back to us many times over. It's a karmic worldview, perhaps, but it's an approach that feels right to me.

Introductions are a different story. Asking for an introduction is different than asking for a referral. A referral is an unsolicited recommendation from a colleague to a prospective client. Like kindness, referrals work best when they happen naturally, but I have no hesitation in asking a friend for an introduction to someone I wish to meet. If I've proven through my work and actions that I am worthy of an introduction, then I have no trouble in asking for one.

The conversation may go something like this example with Jean, a client for whom I have done good work and have established a solid relationship:

Me:	"Jean, I would enjoy meeting your friend Omar, who you serve with on the executive board of Habitat for Humanity. I haven't met Omar, and I would like to get to know him. Would you be willing to introduce us?"
Jean:	"Sure, I'd be happy to do that. Would you like me to send him an email?"
Me:	"That would be great, or I'd be happy to take you both to lunch. Let me know what works best for you two?"
Jean:	"I'll reach out to him and see what his schedule looks like."
Me:	"Thank you, Jean. I really appreciate it. Please let me know if I can do the same for you."

Here are a few rules of thumb that I consider when asking for an introduction:

1. It has to be from someone with whom I have solid relationship (not someone I just met).
2. I have to feel confident that the individual I am asking respects my professional credibility (they have to respect my work).
3. I am not pushy or overly persistent (don't be human spam).
4. I have to accept that the introduction may not occur (if the introduction doesn't occur, I drop it quietly).

I have found that when we invest the time in helping others, most business professionals are willing to do the same for us. I don't give with the expectation of receiving, but I have found that in being considerate and generous to others, we often receive the same treatment. Call it karma, what goes around comes around, or the golden rule – whichever the case, helping others succeed creates the right conditions for a healthy professional ecosystem.

References

David Maister story: Inspired by real characters known by the author, and from David Maister's biography on his personal website: http://about.davidmaister.com/bio/.

Don't Turn into Human Spam: Austin Kleon. *Show Your Work! 10 Ways to Share Your Creativity and Get Discovered*. Workman Publishing, 2014.

Judy Selby quote: Interview by *How to Win Client Business* research team, 2020.

Charles Moren quote: Interview by *How to Win Client Business* research team, 2020.

Jean story: Inspired by real characters known by the author. Names and locations have been changed.

CHAPTER

15

I Can't See the Forest for the Trees

Segmenting Your Ecosystem into Three Distinct Groups

The key to discipline is remembering what we want.
—David Campbell, Canadian politician

Sometimes aspiring rainmakers struggle at first in building their professional ecosystem – not so much the conceptual part, but the doing part. Most professionals quickly grasp the importance of building an ecosystem of individuals who care and support one another. And we understand the value in having 200 individuals who can influence our careers.

The 200-person ecosystem idea is encouraging – it's far less intimidating than 10,000 or 1 million people. But building relationships with several hundred people is also a bit daunting. It's hard enough to maintain close relationships with a significant other, our family, and a handful of close friends. Relationships take time. We have to nurture our relationships if they are to grow. How am I possibly going to do this with hundreds of people?

One of the secrets to effectively building your ecosystem is to see it not as one whole, but as three distinct groups. The first group is your inner circle; I call this Tier 1. Tier 1 is comprised of the most important people in your ecosystem. For most of us, Tier 1 is made up of a few dozen people, maybe 25 to 50 individuals. These are the ones we are going to focus on and spend the most time with in building a close relationship.

Tier 2 is comprised of individuals who are also important to us, but not to the same degree as our inner circle. They are on the outer edge of Tier 1, and we'll also spend time with these people, but we won't invest the same amount of time and effort into building close one-on-one relationships. For most of us, there will be roughly 50 to 100 individuals in Tier 2.

Tier 3 is comprised of the people who we need to know and who need to know us, but we probably won't have close relationships with them. This group is comprised of somewhere in the order of 100 people whom you will stay in contact with a few times a year.

When we think about our network as three distinct groups, it gives us a framework to plan our daily time commitment to each. Doing so allows us to schedule more time and attention to those closest to us and less time with those further away.

The three distinct ecosystem groups are:

- Tier 1: 25–50 people
- Tier 2: 50–100 people
- Tier 3: 100+ people

The Tree Farm Analogy

My father became a tree farmer in his retirement years. He has about 200 acres of farmland in SC where he grew up. Over the past decade, he has devoted much of his time to nurturing the pine trees that he has planted on his family's farmland. Nearly every week, he spends a day at his farm caring for his trees.

My dad jokes that he'll never see the fruits of his labor – the trees will take 25 years to fully mature. This may be true – the 25-year part – but I think he has enjoyed the fruits of his labor. The enjoyment for him is in the weekly stewardship. He enjoys working on his farm, watching his trees grow each season, and seeing the wildlife flourish among the new pine forest.

I think there is value in thinking about our professional ecosystem as a tree farm and in seeing ourselves as tree farmers. The fruits of our labor may take many years to fully materialize, but there is daily joy in caring for our ecosystem and in watching it grow.

Imagine that you have a 200-acre farm – this is your professional ecosystem. In the center of your land you'll build a homestead – a beautiful farmhouse on 20 acres. Your home will be surrounded by a barn, a workshop, a vegetable garden, and a few acres of fruit trees. This is where you will spend 80% of your days. You'll care for your homestead. There will be well-worn paths between the house, barn, workshop, garden, and fruit trees. Your homestead is your inner circle, or Tier 1 ecosystem.

But there is much more to your tree farm. Surrounding your homestead is about 50 acres of walnut trees that you have planted. There are also paths from your homestead reaching out into the walnut trees, but these paths are less worn. You care for these trees too, but it's not an everyday or weekly activity. Perhaps you venture out into the walnuts every few months to check on things. Maybe the wind has blown down a few limbs or a tree needs a little special attention. Think of these middle 50 acres as your Tier 2 ecosystem.

Finally, we have the outskirts of our tree farm. This part of your farm is made up of about 100 acres of oak trees that you have planted beyond the walnuts. The paths grow thinner still as you walk among the oaks. When was the last time you visited these trees; was it last fall? You do make it a point to care for these trees as well, but you simply don't have enough time in the day to get out here very often. Your focus is on caring for your homestead and the surrounding walnut trees. Your outer 100 acres of oaks is your Tier 3 ecosystem. (See Figure 15.1.)

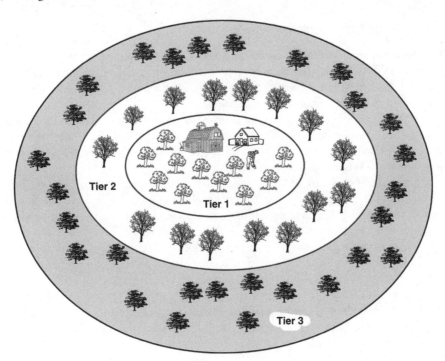

FIGURE 15.1   Your 200-Acre Tree Farm

When we think of our professional relationships as a tree farm, it becomes clearer where we need to devote the majority of our time. Naturally, just like the 20-acre homestead, we'll spend most of our time caring for the few dozen people who will have the greatest impact on our career. We'll connect with these individuals every month of the year. Every few months, we'll check in on our Tier 2 and look for ways to be helpful. Once or twice a year, we connect with our outer circle of colleagues to see what's going on and look for ways to add value.

Caring for Your Professional Ecosystem

As you begin to put the rainmaker skills into practice, start by identifying your Tier 1, 2, and 3 relationships. Begin with a notepad or spreadsheet. Tier 1 is comprised of the key people you either have or want to build a close relationship with. Tier 2 will be your secondary ecosystem partners – important people to know but not as closely as your inner circle. Tier 3 is comprised of the people you'll want to keep in touch with once or twice a year.

Tier 1: Your Inner Circle

An attorney friend recently shared with me that his inner circle encompasses a handful of CPAs and insurance providers. They all serve the same business clients and work together to help their clients succeed. When a new client needs help, they often bring in the others to help solve problems.

The jet aircraft consultant we met earlier includes his previous clients, aircraft management companies, and pilots of corporate aircraft in his Tier 1. He talks with these individuals on a regular basis and they share opportunities. These professionals serve as seeing eye dogs for one another.

Another colleague of mine specializes in strategy consulting within the health sciences field. His Tier 1 is comprised of software companies and related consultants that also serve this industry. He stays in close contact with this inner group of colleagues and keeps each in the loop on the needs of their clients.

For most of us, our current and former clients will be within our Tier 1 relationships. Staying in close touch with those who believe in us and value our work will help open the pathways to repeat business and referrals.

You'll meet many of your closest colleagues naturally in getting involved in the associations, boards, and groups that you share common interests with. In meeting people and making friends organically, we'll find those who share similar beliefs, philosophies, and interests.

Over time our Tier 1 will evolve. People will move on to new companies. Some will change careers. Others will focus on new target audiences that are different than ours. It doesn't mean that we exclude them from our lives, but it may mean that we move them to Tier 2 or Tier 3 and stay in touch quarterly or once a year.

Each day we'll search for ways to stay in touch with our Tier 1, demonstrating that we care. We'll make a habit of taking an hour or two each day to invest in these relationships. We'll follow up with an introduction to another friend. We'll call to say thank you. We'll have coffee to check in to see how they are doing. We'll chat at a board meeting. We'll call them on their birthday.

Tier 2: Your Secondary Circle

Your Tier 2 will be the people who are of secondary importance in your professional ecosystem. It may include members of a trade association, or people who work in adjacent fields. If you're an architect, maybe it's a real estate professional or appraiser. If you're a branding consultant, maybe it's the owner of a branded gift company or a brochure printer.

Because relationships take a commitment of your time, you'll keep in touch with your Tier 2 less frequently and in less personal ways. You may meet your Tier 2 at semi-annual conferences and other related meetings. Perhaps you invite them to join your monthly email newsletter or blog. If you read a good article, you'll clip it and send it to them. You'll congratulate them if you see them recognized in a trade magazine. And you'll call if you hear about an opportunity that may be a good fit for them.

Eugene Buff is an innovation consultant in Boston specializing in commercialization, licensing, and tech transfer. Eugene Buff provides a perfect example of how to stay in touch with your Tier 2 network. Eugene shared that it's very important for him to attend his industry association's meetings twice each year. He frequently speaks at these events (demonstrating his professional expertise), but the biggest benefit, he says, "is keeping in touch with his wide circle of colleagues. If I don't attend, I'll hear from friends asking, 'Eugene, where were you?'" Clearly, Eugene has worked hard over the past 25 years in building his professional ecosystem.

Sometimes individuals will flow between Tier 1 and Tier 2 over time. As a relationship develops, you'll have an intuitive sense of whether this person is someone you need to speak with every month, or stay in touch with once a quarter.

Tier 3: Your Outer Circle

In a well-known sociology study at Stanford in the early 1970s, Mark Granovetter tested the common assumption that we get the most help from our closest ties – our Tier 1. Included in his study were individuals from managerial and technical professions who had recently changed jobs.

What Granovetter found was surprising. He discovered that professionals are nearly two times more likely to find a new job from an acquaintance than from our closest colleagues. Nearly 17% of the new jobs came about through the professional's Tier 1 network, while 28% came from those they didn't know very well – a Tier 3 connection.

Gronovetter's findings were published in the *American Journal of Sociology* and titled "The Strength of Weak Ties." His paper has become a well-cited classic over the past 50 years. The paper highlights the fact that our Tier 1 connections tend to travel in similar social circles. Thus, our closest ties tend to know many of the same opportunities as we do. Our Tier 3, however, serves as a bridge to new information. Our weakest ties are more likely to open up access to new professional ecosystems – facilitating the discovery of new information.

The outer circle is comprised of your third-tier network of professionals. These are not your closest friends and colleagues, but rather your acquaintances. It could be vendors of related products and services, or it could be someone you served on a board with. It may be a client that you did good work for many years ago, or a colleague you worked with early in your career. Perhaps it is a friend from college.

There is value in staying in touch with your Tier 3, but it won't be a daily or monthly commitment. We'll share a cup of coffee with them at an annual conference, or send them a white paper that we've written. We'll stay in touch through LinkedIn, congratulating them on a work anniversary. We'll include them in our newsletters. We'll help them when an opportunity arises. We'll be thoughtful in sharing information when it comes our way, and look to them for advice as needed. We'll build our Tier 3 connections – because, as Granovettor discovered, there is strength in our weak ties.

References

Tree Farm Analogy: Inspired by real characters known by the author.

Eugene Buff quote: Interview by *How to Win Client Business* research team, 2020.

The Strength of Weak Ties: Mark Granovetter. "The Strength of Weak Ties: A Network Theory Revisited." *Sociological Theory* 1 (1983): 201–233.

CHAPTER

16

Why Advertising Doesn't Work for Us

Leveraging Your Firm's Brand Reputation and What to Do When You Don't Have One

The secret to success in any business is to deliver a great, compelling product. No amount of marketing savvy, salesmanship, or operational excellence can overcome a weak product.

—Michael Hyatt, author, *Platform: Getting Noticed in a Noisy World*

I received a call yesterday from an exasperated CEO of a product design and engineering firm headquartered in the Midwest. The firm is 20 years old and includes many blue-chip companies among its clients.

CEO: "I'm at my wits end with advertising."

Me: "I can tell you're frustrated."

CEO: "We've tried everything I know to do. I have hired two of the best advertising firms in Chicago, and have spent hundreds of thousands on advertising. We've rebranded our company, redesigned our website, and launched a PR and promotion campaign."

Me:	"Did you see much improvement in your lead generation pipeline?"
CEO:	"Not one bit. We haven't earned a dime on the money we've invested in any of our advertising."
Me:	"Well, you're doing something right – you've got some great clients and have built a successful company. Where does your business come from?"
CEO:	"Our work is practically 100% repeat business and referrals. We have great clients that have been with us since Day 1. The others have been word of mouth."
Me:	"That's impressive. What is your biggest challenge?"
CEO:	"I don't want to have to rely on repeat business and referrals. I want our brand to be known. It would be great if the phone would ring because people know that we do great work. It would be very inspirational for my team to see that others recognize that we are doing great work."
Me:	"What's wrong with repeat business and referrals? That seems like it's working well for you."
CEO:	"I'd like to grow faster. Our goal is to double in the next five years. I think we need more inbound inquiries from prospective clients in order to achieve this. You would think that an advertising firm would know how to do this."

I empathized with this CEO's pain. I, too, once spent heavily on advertising my company's brand. Like this frustrated CEO, I also came to realize that advertising didn't work in growing our firm's revenue. Unfortunately, we both learned the hard way after a significant withdrawal of our company's cash resources.

Recognizing the Limits of Advertising

If we're in the market for a new smart phone and we see an ad for the new iPhone 11 – sure, we may run over to the Apple Store and try one. Unfortunately, advertising won't accomplish this for you and me. Whether direct email, banner ads, billboards, magazine ads, TV ads – you name it – these approaches won't sell your services in the same way it will for products.

There is some value in advertising, but the benefits are not what you might imagine. And for most of us – unless we're Merrill Lynch or Accenture – we can't afford it. According to research by the advertising industry, roughly $1 billion is spent on advertising in the U.S. each business day. (That's one billion with a B.)

If you were to spend $1 million in advertising annually to promote your firm's brand, that amounts to less than $5,000 each day. Spending $5K won't get you much attention in a noisy world deluged with $1B in daily advertising. Your money will disappear faster than a single drop of rain in the Sonoran Desert, with about the same impact. Besides, most of us can't and won't spend $1M on advertising.

And even if we could afford it, I'd say the money would be better spent in other ways. In our world, advertising does one thing: it can bring awareness to your firm, but it's an expensive tool.

After years of advertising, a prospective client might say to you, "Yeah, I've heard of you guys before. Tell me again, what is it you do?" That's about it. Advertising won't cause a prospective client to pick up the phone and call your front desk to say, "Hey, I'd like to buy some consulting. Can you put me through to the person in charge of that department, please?"

Advertising can't earn you respect and trust. This can only be built one day at time, one relationship at a time, in doing great work for clients and demonstrating your expertise to the world. Whatever promotional approaches you dream up won't make the cash register ring.

When was the last time you clicked on a banner ad? How many times this week have you pored through your emails, credit card in hand, cruising to find a good HR consultant or M&A advisor? It may work for selling Caribbean cruises, but for us this is pure fantasy; far better to invest the time in doing the things that work than wasting our precious time pursuing quixotic dreams.

This may be alarming news if you've just become the chief marketing officer of a services firm. If you're a new CMO and have a $1M advertising budget, then go ahead and spend it like a drunken sailor and enjoy every minute of it. Just don't delude yourself – or the firm's partners – into thinking that it will grow revenue. It won't.

A Better Way: Building Relationships

There are only two reliable things you can do as CMO to help your firm grow: help your people meet others naturally, and create opportunities for your firm's people to demonstrate their expertise. These are the only two activities I've witnessed that will consistently help you build relationships with those you wish to serve and earn respect for your team's professional capabilities.

- **Meet People**: Create or participate in events and organizations (forums, conferences, seminars, boards, charities, etc.) that give your people an opportunity to build their professional ecosystem
- **Build Content**: Assist your people in creating good content (papers, research studies, books, podcasts, seminars, speeches, etc.), which demonstrates your expertise, and share this information widely with those you wish to serve

Whether you're a chief marketing officer, an aspiring partner at a well-known firm, or a solo professional, these two topics are worthy of a closer look, because, unlike advertising, these two approaches initiate opportunities for real relationships with actual people who could use your help.

Approach 1. Meeting People: Opportunities for Making Friends in a Natural Way

Building off of the playground analogy, we make friends more effectively when we share mutual interests. Therefore, rather than spending money on advertising campaigns, I would look for opportunities to join, attend, and host events that are related to your professional and personal interests.

Advertising is so tempting. It's so much easier to spend tens or hundreds of thousands on advertising than it is to build real relationships. There is also a touch of ego in advertising as well. When we see our firm's name spread across a huge stadium billboard, our chest swells with pride. Once again, if you've got money burning a hole in your pocket, go for it. Just don't sit back waiting for the phone to ring.

If your aim is to actually win more client business, then a better approach is to make friends with the 200 people you need to know. And making friends naturally at events, seminars, charities, and conferences is a great way to do so.

You may recall Mike McCracken, the jet aircraft broker we met earlier. For decades he sold jets to Fortune 500 private companies and to high-net-worth individuals. He spent the bulk of his successful career as a salesman for well-known aircraft brands like Beechcraft and Embraer. And he was one of the best at his job: knowledgeable and highly professional.

When I first met Mike he had recently gone out on his own as a jet aircraft consultant. His goal was to help others avoid the costly mistakes many first-time jet customers make. No longer was he working for a company with a well-known brand; it was just him and his decades of industry knowledge.

Our conversation went something like this:

"Tell me more about your approach to meeting those who could use your help," I prompted.

"Well, I've been sponsoring a golf tournament in Florida," Mike offered.

"How's that working for you?"

"I haven't gotten a call yet from the golf sponsorship. Can't say it's doing any good – near as I can tell," he responded.

"Where have your clients come from so far?" I asked.

"Well, let me think," Mike reflected. "One was a former Fortune 500 CEO I served for years at Beechcraft. Another was a referral from a client I recently helped buy a company jet. The most recent was an introduction from a charter pilot I caught up with at an industry conference."

I smiled to myself as Mike candidly shared his story with me. I'm sure you may be smiling as well in seeing the pattern in these stories.

"Supposing you took the advertising money from the golf tournament and reallocated that money to travel – attending more events and visiting with people you've worked with in the past? Perhaps that would be a better way to get more clients," I suggested.

"Yeah, I see your point. What you're proposing is just a lot more time and effort," Mike reflected. "I was hoping to dial back my travel time. I thought maybe some advertising would help me get some more business."

Mike has a great reputation with the people who have known him for years. He also has an incredible wealth of industry knowledge that is worth a lot to people interested in buying or selling a jet aircraft. But knowing what we now know about the client's buying decision journey, advertising isn't likely to help Mike build his practice.

Furthermore, advertising is an expensive use of Mike's hard-earned money. A better approach is to look for opportunities to connect and reconnect with people we can serve. And in building our ecosystems, we're reallocating our limited advertising budgets toward approaches that will actually lead to new client business.

Approach 2. Building Content: Opportunities for Demonstrating Our Expertise

Creating good content is an effective approach to initiating conversations with people who could use our expertise. Providing insightful content serves two rainmaker purposes: it demonstrates our expertise and builds our professional ecosystem.

If fact, while advertising won't inspire a prospective client to pick up the phone and call you, good content can. Content can generate inbound inquiries in a way that advertising does for consumer products. It's the closest parallel to advertising that exists in professional services.

When you write a good article and share it, that provides a natural means for others to find you. And, unlike advertising, it establishes you as a person with valuable insight. Similarly, when you give a good talk at an industry event, it is a great way to connect with others who resonate with your message.

Content marketing has become an industry unto itself. And for good reason: it works. One of the industry's leaders, Hinge, specializes in helping those in professional services build a following through well-designed content. According to Lee Frederiksen, managing partner at Hinge:

> Powerful content helps your audience understand who you are, and what your approach is to things. It is not self-serving or sales copy, but content which educates them, informs them, and helps them understand the challenges they may be facing.

If someone needs help with an important problem, they will search for answers on the internet. This may not be their only approach to finding help – they may also lean on trusted colleagues for advice – but they will enhance their search by seeking out good information on the web.

If you've written a good book on the topic or hosted a related podcast, people in search of help will find you. And if they share your point of view on the issues and are looking for help, they will often reach out to you. Building good content is an effective way for us to demonstrate our expertise and build our professional ecosystem.

Content is simply a natural starting point for a discussion. And from these early conversations, true relationships can emerge with people who need your expertise. Creating and sharing content requires a commitment of your time. I wish we could wave a magic wand and make good content appear. Unfortunately, we know good content takes time to develop.

Content has to be created by those with expertise. Others can assist in packaging an expert's knowledge, but it's still going to take a fair bit of heavy lifting on our part to get it created. We can't outsource or delegate the knowledge in our head.

As we've discovered, cold-calling isn't fun, and it's not highly effective for most of us. Advertising can be fun, but unfortunately that isn't very effective either. The two best approaches to building your ecosystem are through rubbing elbows with your target audience and creating good content for them. All other approaches are simply not very effective. At the end of the day, the only people who will ever hire us need to come to know, respect, and trust us. And this brings us to the next rainmaker skill: *developing trust-based relationships*.

Leveraging Your Firm's Brand

If you work for a prestigious firm that is among the most well-known and respected in your industry, then you are more likely to receive a favorable reply to a cold call or letter. If your employer is someone like Goldman Sachs, Deloitte, or McKinsey, your firm's name carries a lot of weight.

If you're a VP at Citibank and you get a call from a partner at Goldman Sachs inviting you to lunch, there's a reasonably good chance you'll accept this invitation. If you're a partner at General Financial Services (or any other unknown brand) and you try this same approach with the VP at Citibank, the odds are you'll hear nothing. This has been my experience and is supported by countless stories from others.

I recently spoke with Howard Hull, the group tax director for the Al-Futtaim Group based in Dubai. The Al-Futtaim Group is a family-owned company with 42,000 employees in 31 countries. Their wide-ranging businesses include automotive, retail, real estate, healthcare, and financial services.

Before joining Al-Futtaim, Howard was an international tax partner at EY (Ernst & Young) for ten years, with posts in Geneva, London, and Dubai. "The EY brand helped enormously," Howard said. "The brand was very important in getting in the door." Yet Howard quickly added this caveat: "But the EY name wouldn't win you the business. You had to establish your personal value and build trust. If you couldn't do that, the client's business would go to someone else."

Those who work for a prestigious firm have a leg up in getting invited into the C-suite. Strong brands built on decades of great work will open doors for you. Those fortunate to be in this position have the hard work of former partners to thank for this. A firm's reputation will only get you in the door, though. From there on, it's up to you to demonstrate your expertise and build a relationship with the prospective client; the great firm name will only carry us so far. For the rest of us – and I'm guessing this is most of us – we'll have to use different approaches in building our professional ecosystem because cold-calling will not be nearly as effective for us.

References

CEO story: Inspired by real characters known by the author. Names and locations have been changed.

U.S. advertising spending data: A. Guttmann. "Media Advertising Spending in the United States from 2015 to 2022." *Statista*, 2019.

Mike McCracken story: Interview by *How to Win Client Business* research team, 2020.

Lee Frederiksen quote: Lee Frederiksen. "Top 10 Marketing Techniques for Professional Services." Hinge Marketing, www.hingemarketing.com.

Howard Hull story: Interview by *How to Win Client Business* research team, 2020.

Skill 4: Develop Trust-Based Relationships

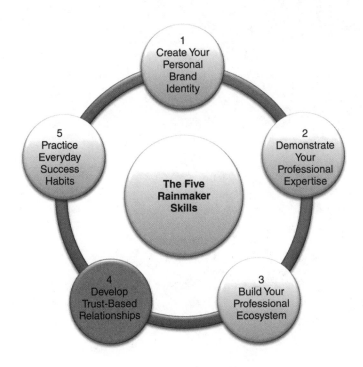

17

What Is Trust and Where Does It Come From?

Do Clients Really Hire People They Like?

Trust is a living, breathing, emotional bond that connects people to one another. It's intimate, personal and powerful. In a world where it seems like everyone is out to pitch, scam, or screw you, it is also a rare and precious commodity.

—John Hall, author, *Top of Mind*

Tom McMakin and I stirred up a bit of controversy in *How Clients Buy* when we claimed that respect and trust were more important than being liked. Our claim went against conventional wisdom because many believe clients hire people they like.

We suggested that likeability is a tiebreaker in the client's decision-making journey. Between two equally qualified and trustworthy candidates, a client would give the nod to the one they liked better. We proposed that respect and trust were more important than being likeable: respect for our professional ability and trust that we had the client's best interests at heart.

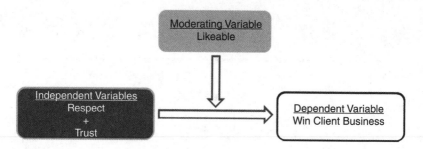

FIGURE 17.1 Being Likeable Allows Respect and Trust to Grow

As I've studied this more carefully, I've come to believe that likeability is what social scientists call a moderating variable. A moderating variable (in this case being likeable) influences the relationship between the independent variables (respect and trust) and the dependent variable (you being selected or not). (See Figure 17.1.)

Or, put another way, being likeable on its own will not win you client business, but not being likeable can lose you client business. I have come to believe that being likeable allows respect and trust to begin to grow in a relationship.

Recognizing Authentic Likeability

Perhaps the controversy we created stems from differing opinions of what it means to be likeable. What makes a person likeable anyway?

Here are a few qualities that quickly come to my mind. Likeable people:

- Are interested in you
- Ask good questions
- Are good listeners
- Are helpful
- Are comfortable with who they are
- Are informative – they share interesting information
- Are honest

Alternatively, unlikeable people:

- Are self-centered and egotistical
- Talk too much about themselves
- Don't seem to care about you
- Are more interested in getting than giving

- Exaggerate their accomplishments
- Try too hard to make a good impression

We've all met people who fit the description of the likeable person. They make us feel good. We learn from them and they inspire us. They are honest and look for ways to help us. They are upbeat and pleasant to be around.

We've also met people who were not very likeable. They talk a lot about themselves but don't seem that interested in us. They seem most interested in what they might get from us. They aren't very enjoyable to be around. They leave us feeling drained.

Your list of likeable traits may be different than mine, but I bet there is considerable overlap. I wonder if being likeable is really about showing others that we care about them. When we listen closely to what someone is saying, aren't we demonstrating that we care? Caring and empathy make us more likeable, right?

I think the risk we run in trying to be likeable is that we try too hard to make a good first impression. In her best-selling book *Presence*, Harvard psychologist Amy Cuddy offers this perspective:

> When we are trying to manage the impression we're making on others, we're choreographing ourselves in an unnatural way. This is hard work, and we don't have the cognitive and emotional bandwidth to do it well. The result is that we come across as fake.

Trying to be more likeable feels phony. Striving to be more thoughtful doesn't. And when we show others that we care about them, we are building trust.

Where Trust Comes From

I reached out to a good friend who has had a long, successful career in business. I posed to him the following question: think for a minute about two people from your business career whom you trust the most. For each of these individuals, answer the following.

1. Why do you trust them?
2. What specific behaviors do they demonstrate that makes you trust them?

Here was my friend's reply:

Person 1: Honest, ethical, does what is right whether it helps him or not, concern to do the right thing, consistent and predictable. I can predict based upon his character what he will do in a situation

Person 2: Honest, ethical, does what he says he will do. Will do the right thing even if it means less money for him.

Then, I asked my friend to identify the common themes that stand out. Here was his reply:

- Honest
- Ethical
- Reliable
- Truly believes in win/win
- Does the right thing independent of financial incentives

Next, I asked my friend this: Think about two individuals whom you trust the least in business:

1. Why do you *not* trust them?
2. What specific behaviors do they demonstrate that makes you *not* trust them?

Person 3: Always thinks about what is in it for him, competitive to the point of winning is better than an ethical solution, has demonstrated it is all about his win numerous times over the years

Person 4: Thinks about himself first, not a win/win guy, things he says about customers make me feel like they are a meal ticket, not a long-term client. There is just something about him that makes you think to yourself, "watch out."

Here are the common themes of those my friend trusted the least:

- They come first
- Not honest, hide their real agenda
- Ethical when it is convenient
- Will do anything to win

I bet if you tried this exercise your answers wouldn't be that different from my friend's. There seems to be some commonality between my likeable traits and my friend's trustworthy traits.

I have come to believe there is an interesting connection between being genuinely likeable – in the ways described above – and building trust. Trust begins when others sense that you care for them. When we demonstrate that we have the other's best interests at heart, we engender trust and, in an odd twist, we become more likeable.

Maybe the old adage that people hire people they like is true. Maybe it's just a definitional hang-up for me. If you were to tell me that caring is likeable and showing empathy for others is at the heart of building trust, then perhaps we're getting close to the foundation of successful client relationships. Some may call it being likeable; I'll call it being less self-centered and more caring toward others.

Building trust-based relationships is vital to our success as rainmakers. And showing empathy for others is at the core of building trust. Thus, in order for us to succeed we have to first demonstrate that we care. We have to put our client's interests first. We have to make them the focus of our attention. We have to talk less about ourselves and listen more to their stories. And when we care about others first, we become more successful ourselves.

One of the most respected professional services firms in the world is the management consulting firm McKinsey & Co. Looking closely at McKinsey's values reveals some interesting patterns as it relates to the themes of empathy and honesty in building trust-based relationships:

- Put client interests ahead of the firm's
- Observe high ethical standards
- Preserve client confidences
- Build enduring relationships based on trust

For many companies, value statements are simply platitudes posted on a breakroom wall. I have a handful of friends who have worked at McKinsey, and each confirms that their values are real – something that guides their actions with clients every day. I believe this is a major part of McKinsey's strong reputation and success.

We live in a world where trust is at nearly all-time lows. According to a 2018 report by the communications marketing firm Edelman:

> Only a third of Americans now trust their government "to do what is right" – a decline of 14 percentage points from last year. Forty-two percent trusts the media, relative to 47% a year ago. Trust in business and non-governmental organizations, while somewhat higher than trust in government and the media, decreased by ten and nine percentage points, respectively. Edelman, which for 18 years has been asking people around the world about their level of trust in various institutions, has never before recorded such steep drops in trust in the United States.

Put others first, listen to them, be honest, and do the right thing. Is it really that simple? If this is true, isn't it rather strange that it's not very common? If trust were common, we'd see these values in action every day. Because this behavior is not common, it sets those apart who live these timeless principles in their relationships with others. It may be simple, but apparently it's not easy, which is why relationship-builders are coveted by every firm in any profession.

Advice from a Seasoned Relationship Builder

Soliciting college donations may be one of the most challenging professional jobs. At the end of the day, what is the benefit of someone giving a major bequest to a college – a warm fuzzy feeling and a family legacy? Fuzzy feelings and legacies are hard to quantify, and there is no shortage of things a donor could do with a million dollars.

Coker College is a private, liberal-arts college based in Hartsville, South Carolina. Most of us haven't heard of Coker or Hartsville; it's a small college (now Coker University as of the fall of 2018) in a small town in a rural corner of the state. But it happens to be my hometown. And I happen to be family friends with the man who devoted nearly 40 years to raising money for Coker College. His name is Frank Bush.

When Frank arrived at Coker College in 1975 he was barely 30 years old. He spent the first years of his career in admissions at Limestone College and then at the College of Charleston. Frank came to Coker as head of admissions – a big job for a young man not even 10 years out of college.

Four years after Frank's arrival in Hartsville, Coker's president was searching for a new development director – the person responsible for soliciting donor funds for the college. After a lengthy search for the right person, someone on the board of trustees suggested, "What about Frank Bush? Everybody likes Frank."

"I don't know anything about raising money," Frank replied when the college president floated the idea by him. "You can do it," the president insisted; "Here's a list of ten people you need to visit with. By the way, your goal is to raise $130,000 this year. Good luck!"

And that's the way Frank got started in college fundraising. He went on to raise millions upon millions of dollars for Coker over the coming decades – adding new libraries, art centers, sports facilities, classrooms, and dorms. If you're ever in South Carolina, you should visit the Coker College campus. It's as lovely a campus as you'll find anywhere. And the progress that Coker College has made since the 1970s is tied to the fundraising success of Frank Bush.

First on the list of people Frank was to visit was the CEO of Sonoco Products Co. It was a short commute because Sonoco's headquarters were a few blocks from campus. The CEO's name was Charlie Coker – his grandfather had founded Coker College in 1908.

After exchanging pleasantries, Frank got straight to the point. "Mr. Coker, I'd like to ask you to make a gift to Coker College's unrestricted annual fund for $1,000."

"I've never personally given a gift of $1,000 to Coker College before, Frank. That's a lot of money," Charlie replied.

Frank was taken back but he persisted. "Well, Mr. Coker, I'm going to have to be candid with you. Coker College needs for you to make this gift of $1,000, because if you don't the college will not succeed and the fundraising program under my direction will fail. And if I can't be successful, then I'm going to have to go back to the president today and resign from this important position."

After a long pause, Charlie asked his secretary, Elsie Byrd, "Would you please bring me my checkbook?" Mr. Coker proceeded to write the college a check for $2,000. And Frank went on to successfully raise $130,000 in Coker's first annual fund campaign.

With the help of Charlie Coker and many others, Frank went on to a long, successful career at the college, eventually rising to become its executive vice president, responsible for all donor development and student admissions activities. Along the way he built a reputation as an influential leader in his profession and served on the board of directors of the Council for Advancement and Support of Education (CASE) – the global nonprofit association dedicated to educational advancement. Upon his retirement in 2007, Frank was asked to address 2,000 members of CASE at their annual conference in New York.

I asked Frank what he thought were the keys to his success in fundraising. Here's what he shared with me:

1. It's all about relationships.

2. Always be yourself; don't pretend to be someone you're not.

3. Do not misrepresent yourself. Don't exaggerate. Tell the truth. When you make a mistake, apologize.

4. Be respectful, but don't be intimidated by important people – they are people just like you and me.

5. Be observant and listen. Let people tell their story. In their stories you'll find their desires, motives, values, hopes, and dreams.

6. Do your homework; know the people you're calling on and find a way to connect with them personally.

7. Don't ask for money. It's not about the money; it's about what the money can accomplish and why the mission is important.

8. You have to believe in your product. If you don't, you won't succeed.

9. If you know you have the best product and you didn't win the customer's business, then you know you did something wrong. Learn from your mistakes and how can you do better next time. Don't take "No" personally.

10. The best donor is a past donor. Value each donor independent of how big or small their gift. Each gives according to their abilities.

I believe that Frank's advice will serve each of us as we work to build trust-based relationships with those we wish to serve. There are several interesting themes woven between Frank's advice and the traits we discussed earlier. Two of the biggest are empathy and honesty.

- When we listen to someone, we demonstrate empathy and build trust.
- When we're honest with people and true to our authentic selves, it engenders trust. It's hard to trust someone who is dishonest or phony with you.

Trust is an essential component of the client's buying decision journey. Because trust is so important in winning client business, all successful rainmakers demonstrate the ability to earn it with those they wish to serve.

References

Amy Cuddy quote: Amy Cuddy. *Presence: Bringing Your Boldest Self to Your Biggest Challenges*. Little, Brown and Company, 2015.

Where Trust Comes From story: Inspired by real characters known by the author. Names and locations have been changed.

McKinsey & Company Values: "Our Purpose, Mission and Values," www .mckinsey.com.

Edelman trust survey: Uri Friedman., "Trust Is Collapsing in America." *Atlantic*, January 21, 2018.

Frank Bush story: Interview by *How to Win Client Business* research team, 2020.

18

Conversation Skills for Introverts (and the Rest of Us, Too)

Using Small Talk to Find Common Ground

The most basic and powerful way to connect to another person is to listen. Perhaps the most important thing we ever give each other is our attention. When people are talking, there's no need to do anything but receive them. Just take them in. Listen to what they're saying.

—Rachel Naomi Remen, MD, author, *Kitchen Table Wisdom*

People feel comfortable being around people with whom they have things in common. Remember the playground idea I shared earlier – kids tend to hang out with kids who enjoy the same activities as they do. This analogy applies to finding common ground with people in business. Donald Miller discusses the importance of common ground in his *New York Times* bestselling book *Building a Story Brand*.

> Customers look for brands they have something in common with. The human brain likes to conserve calories, and so when a customer realizes they have a lot in common with a brand, they fill in all the unknown nuances with trust. Essentially, the customer batches their thinking, meaning they're thinking in "chunks" rather than details. Commonality, whether taste in music or shared values, is a powerful marketing tool.

Miller's insight applies to our personal brands as well. This is why so-called "small talk" is so important – the seemingly irrelevant chit chat that often begins a business conversation. Small talk creates an opportunity to find common ground, and to begin to get to know one another on a more personal level.

In many parts of the U.S. and the world, it's perceived as rude to get right down to business. The length of the small talk varies from region to region. In some parts of the world, small talk may go on for hours or days. In the southern U.S., small talk may last 15 minutes; in New York, maybe five. I have come to realize there is significant relationship-building value in small talk. It's important to be sensitive to cultural norms, but the benefit that it provides is significant.

One of Frank Bush's relationship tips was to know the people you're calling on and find a way to connect with them personally. Connecting on a personal level is finding common ground. When we find that we have things in common with someone, we begin to connect on a more human level and the early glimpses of trust begin to emerge. As humans, we tend to trust people more when we have things in common.

Many of us in professional services tend to be more introverted. Some of us reluctantly admit that talking with new people is challenging. And while we don't have to be the next Oprah Winfrey or Stephen Colbert, I do think there is value is learning to find common ground with others through small talk. Perhaps the best news for us introverts is that good listeners are often the best conversationalists.

Adam Grant, Wharton professor and author of *Give and Take*, offers this point from his research: "the most successful relationship builders frequently ask thoughtful questions and listen with remarkable patience." This is a skill that lends itself well to introverts.

Small talk also lends itself to those who are naturally curious; curiosity about new things leads to learning and listening. Murray Joslin, senior vice president at Integreon, shared that "We have found over the years that innate curiosity is one of the keys to business development success. Curious people want to learn and listen – skills that lead to success in building relationships."

Amy Cuddy – the Harvard psychologist we met earlier, studied the role of listening in how others perceive us. Her research led to an interesting finding she refers to as the paradox of listening. Rather than speaking, asserting, and knowing, we become more powerful when we stop talking, stop preaching, and listen. According to Cuddy, here's what happens:

- People can trust you.
- You begin to see other people as individuals, and maybe even allies.
- You acquire useful information.
- You develop solutions that other people are willing to accept and even adopt.
- When people feel heard, they are more willing to listen.

There are an infinite number of ways to find common ground with other people: current events, sports, hobbies, schools and colleges, regional culture, kids, employers, and travel – just to name a few. With social media, we can quickly learn a lot about people. We can see the cities where people live, what they do professionally, who they work for, where they went to college, and often what their hobbies are. In a few minutes before a phone call or a meeting, we will discover numerous ways to find common ground with a new friend or colleague.

We can also learn more about someone through a mutual acquaintance, hence the power of our professional ecosystem. We might say to a colleague, "Hey, I'm meeting with your friend, Tamara, later today. What can you tell me about her?"

Doing research on a person's background is a form of listening. And, like listening, researching a person's background shows that we care about them. When we show up already knowing a few things about the other person, it signals that we want to know more about them. When we take an interest in another person, it makes them feel good about themselves. And it makes us more likeable – not in a phony kind of way, but in a sincere, empathetic way. It provides opportunities for finding a common connection.

Finding Common Ground through Small Talk

When I was in my 20s, I moved a lot with my work – 11 times in 10 years. That's a lot of boxing and unboxing. I also traveled a lot for work. For several years I was on a plane three out of four weeks of every month. The moving and travel gave me the opportunity to see every part of the U.S. and many parts of the world.

I learned what it's like in New England, Southern California, the South, and the Midwest. I have traveled through Appalachia, the Northwest, and the mid-Atlantic regions of our country. I worked in big cities – Boston, Atlanta, Chicago, and Los Angeles – and in small towns in Mississippi, Illinois, New York, and Michigan. I also worked and traveled in Europe, Canada, Australia, and New Zealand. That's a lot of frequent flier miles and Interstate rest stops.

The benefit of all of this travel is that it serves as a great source of common ground. In small talk, I might say, "Hey, is that awesome sandwich shop still up by the state courthouse?" or "I remember the first ball game I attended at Fenway Park – Ortiz hit a walk-off homer in the ninth to beat the Yankees."

These little things seem so irrelevant. And yet they are important. They help us find common ground with someone we're getting to know. If a new acquaintance is into something that I'm not, I'll ask questions. "Hey, I know you're a connoisseur of good food. What's your favorite restaurant in San Francisco?" or "What's the best show running on Broadway right now?"

It's not necessary to have moved 11 times in 10 years or traveled like a long-haul trucker.

People love talking about things they are interested in. And they love sharing their knowledge with others. You don't have to be an expert in everything to engage in small talk; just ask good questions and listen. Listening is showing empathy and respect. And, as we also learned from Frank, in people's stories they will share with us their hopes, dreams, interests, and aspirations.

Tim Hartland is an independent financial advisor in Wichita, Kansas. I met him for the first time last Friday on a Zoom video call. Tim reached out to me several weeks ago to say he had enjoyed *How Clients Buy*, and had bought several extra copies to share with colleagues. He wanted to see if I'd be open to having a phone call to discuss several ideas he had about his financial practice.

Naturally, I was happy to talk with Tim. Anyone who sincerely compliments our work makes us feel good about ourselves, and we're glad to spend time with them. When our one-hour conversation began, we quickly discovered we had a lot in common. So much so that our small talk lasted 20 minutes.

As it turned out, Tim played college baseball at Baylor. He played catcher and several of his teammates were now in the major leagues. I love baseball as well. Growing up, I lived with a ball glove on my hand most of the day. Unfortunately, I wasn't especially good at the game, and certainly didn't receive any calls from college recruiters. But our mutual love of baseball gave us a common ground that led to a fun discussion.

It also turned out that Tim and I are both guitar enthusiasts. Our conversation went something like this:

Me:	"What are your interests outside of work and family?"
Tim:	"I enjoy playing guitar."
Me:	"That's cool – me, too. What kind of guitar do you play?"
Tim:	"I own a couple of Martin acoustics."
Me:	"No way, I own a Martin acoustic as well. I play before bedtime to wind down. I'm just good enough to know I'm not very good. But I love playing and I enjoy listening to music."
Tim:	"Me, too. I'm not going to quit my day job as a financial planner, but I do enjoy playing guitar with my friends."

By the time we settled into our business conversation, I felt like I had known Tim for much longer than a few minutes. We had a number of similar interests, and I knew a little about his young kids – I could see on Zoom that his office whiteboard was filled with the kids' art that every parent adores. I shared with Tim that my kids were now in college, but I cherished the memories of my kids' drawings.

While Tim and I have only spent an hour together, I felt an instant connection with him. None of this would have occurred if it hadn't been for our small talk. We went on to have a detailed discussion about his work, and explored opportunities for helping one another. We agreed to keep in touch. I'm already looking forward to our next conversation.

Building Common Ground over Time

Breaking the ice through small talk demonstrates that we're much more than just our professional selves. As relationships begin to build over time, we can return to common interests to pick up where we left off. Like a good sitcom plot that reappears in future episodes, common interests serve as a starting point on subsequent calls or meetings.

I'll share with you a conversation I had this week with a colleague I've known for 25 years. Michael Hinshaw and I share a love of fly-fishing and it serves as a common thread through all of our conversations.

Michael:	"How are you doing?"
Me:	"Great. I got up to an alpine lake over the weekend for some fly-fishing. The rivers are high now with spring runoff, but the high mountain lakes are fishing well."
Michael:	"That's awesome. I've got to get back out to fish again with you this summer."
Me:	"Yes, let's do it. September will be perfect – it's the best month of the year here in Montana."
Michael:	"I'll look at my schedule, but let's make it happen."
Me:	"How are your kids doing?"
Michael:	"Great. My daughter just graduated from college and my son from high school."
Me:	"I know you're proud of them. You should bring them with you when you come to Montana if you can."
Michael:	"Yes, I owe them both a graduation trip."

As Frank Bush shared with us, it's all about relationships. We may be conducting financial audits, providing sales training, or assisting corporations with legal matters, but each one of us is in the relationship business. It's a core piece of the client's buying decision journey.

We all have lives outside of our work – lives full of families, hobbies, issues, and interests. Asking about people's lives outside of work shows that we care about them as humans. And, as we've observed, sincere empathy for others is the starting point of all relationships. As we begin to build relationships with people, finding common ground through small talk serves as a natural starting point. It's something we can all benefit from – introverts as well as extroverts.

References

Donald Miller quote: Donald Miller. *Building a Story Brand: Clarify Your Message So Customers Will Listen*. HarperCollins, 2017.

Adam Grant quote: Adam Grant. *Give and Take: A Revolutionary Approach to Success*. Viking, 2013.

Murray Joslin quote: Interview by *How to Win Client Business* research team, 2020.

Amy Cuddy quote: Amy Cuddy. *Presence: Bringing Your Boldest Self to Your Biggest Challenges*. Little, Brown and Company, 2015.

Tim Hartland story: Interview by *How to Win Client Business* research team, 2020.

Michael Hinshaw story: Interview by *How to Win Client Business* research team, 2020.

19

The Art of Keeping in Touch

Finding Opportunities to Be Thoughtful and Helpful

The deepest principal in human nature is the craving to be appreciated.
—William James, Harvard University professor, often considered the father of
American Psychology

Wayne Baker is a sociologist at the University of Michigan and a leading scholar on the study of professional ecosystems. According to Baker, "If we create networks with the sole intention of *getting something*, we won't succeed. We can't *pursue* the benefits of networks; the benefits ensue from investments in meaningful activities and relationships."

In 2011, *Fortune* magazine recognized Adam Rifkin as the most connected businessperson in the U.S. In a study of LinkedIn connections to the 640 most influential businesspeople, the research found that Rifkin was more connected to the business elite than any other person. When asked about his approach to building his ecosystem, Rifkin offered this:

My network developed little by little, in fact a little every day through small gestures and acts of kindness, over the course of many years – with a desire to make better the lives of the people I'm connected to.

If you think back on the people in your life you trust most, I bet you have known many of these individuals for years, perhaps decades. Trusting relationships are not built overnight, but they can begin on day one. Trust begins when we show others that we care, are interested in them, and want to be helpful.

And when we find common interests, we have a natural avenue to discuss things that matter to us. Small talk allows us to check in with others – to inquire about their lives. And, over time, this kernel of trust begins to grow, like a many-layered coat woven one thread at a time.

In Section Three we explored the importance of building our professional eco-system – the 200 most important people in our professional lives. We learned that our network of trusted colleagues will influence our careers. We discovered that when trust and respect are present, we open up the pathways of repeat business, referrals, and inquiries. And when these pathways begin to flow freely, we aren't forced to practice cold-calling – which we don't like to do and isn't very effective for most of us anyway.

When we make a consistent habit of showing others in our ecosystem that we care about them, we earn trust. And over time, the trust becomes a powerful currency, more valuable than money or gold. With trust, we can accomplish incredible things together. Without it, all things grind to a halt. The upside is so great that it's a wonder these empathetic gestures aren't practiced more commonly. In practicing these acts we stand out in a world where trust is often hard to find.

Finding Opportunities to Be Thoughtful and Helpful

What then do we need to do to demonstrate to others that we care? Here are some thoughtful gestures I've observed in which successful rainmakers keep in touch with others to show they care:

- Recognize other's success
- Provide helpful news/information
- Offer introductions and referrals
- Ask for advice or guidance
- Follow up to inquire about an issue
- Remember important dates

- Give small gifts
- Show gratitude toward others

Let's take a closer look at each of these to see how they work and why they're so effective.

Recognize Other's Success

Herb Richardson is the father of one of my closest friends, David. Herb recently passed away after a good life and a successful career in banking. From Herb I learned an important life lesson: recognize others' success.

When I was in my mid-20s, I had a chance to attend the Kentucky Derby with David. His family has held a box seat at Churchill Downs for generations. The Kentucky Derby is an incredible cultural event. I only attended this one time, but it remains one of the best memories of my life.

Over pre-race cocktails, Herb shared with me a practice that he used to keep in touch with his clients and colleagues. Each Sunday he carefully read the area's largest newspaper: Louisville's *Courier-Journal*. If he saw a news story where a friend, client, or colleague received special recognition, he clipped the article and mailed it on Monday morning with a handwritten note.

Who wouldn't appreciate this small act of kindness? And yet simple gestures like this have faded like the pages of an old newspaper. Herb's act of recognition showed that he cared. And it also served as a reason for staying in touch, and for remaining top of mind to his clients and colleagues.

Perhaps newspapers have been replaced with tablets and laptops, but news is consumed more than ever. Taking a few minutes to print an article and attach a handwritten note would do the same as Herb's newspaper clippings. When we recognize others' success, it's a sincere way for us to keep in touch and show others that we care.

Provide Helpful News and Information

In the spirit of Herb Richardson, we can also do the same with news articles, white papers, or books of interest to a client or colleague. When we come across information of interest and share it with others, it says, "I'm thinking about you."

When we read or watch the news, we can think about how this is relevant to others. It's a genuine way to show others that we care about them. It can be as simple as forwarding a news story via email. Or, better still, we can print the article and mail it with a handwritten note. Or buy an extra copy of a book and mail it to a colleague. And, like Herb's notes, they offer an authentic way to stay present in others' lives.

Offer Introductions and Referrals

Introducing others and giving referrals are two of the most thoughtful things we can do for others in our ecosystem. In connecting people who could benefit from knowing one another, we demonstrate that we care about their success. One of the benefits of having a strong ecosystem is that as it grows, we can connect our ecosystem to others.

Hardly anything feels better to us than when a colleague or client offers us a referral for new business. Unsolicited referrals are the ultimate show of the respect and trust we have for others. After repeat business, there is no better client pathway than a referral. When we refer a close colleague to a prospective client, our faith in someone becomes a source of social currency, a signal that this person can also be trusted.

Offering introductions and giving referrals reinforces the relationship we have with others in our professional lives. We stay relevant and remain top of mind. We show others that we care and seek to help others. It's one of the best ways of keeping in touch.

Ask for Advice or Guidance

When we ask others for advice, it shows that we respect their opinion. If someone calls us up asking for help on a topic, it's the sincerest form of flattery. It shows that others value your knowledge and expertise. I've never met someone who didn't gladly share advice with others they are close to.

When faced with an unfamiliar challenge, think about those in your network who can provide guidance. Asking for advice serves as an organic way to keep in touch, and also sends a message of trust and respect.

It could be a topic of professional interest – for example, "Do you know of a good business attorney in Boston?" Or it may be something relating to a personal interest: "Hey, didn't you go to Hungary last year on vacation? Can you recommend a good hotel in Budapest?"

Either way, people enjoy being seen as an expert. And it provides a natural bridge to staying in contact with one another. And, the interesting thing is, when you ask others for advice, they often seek your advice as well. Such is the nature of mutual trust and respect.

Follow Up to Inquire about an Issue

When you follow up with a client, it signals that you care. Perhaps in conversation over lunch one day a client offers that they are facing a challenge at work or in their personal lives. Maybe it's something relevant to your work – or maybe not.

Maybe your colleague is seeking guidance on possible software vendors for their business. You make an introduction to a company you worked with in the past. Then, a few weeks later you follow up to see how the initial meeting went.

Or maybe your colleague is looking for a good soccer camp for their daughter for the upcoming summer. Your own kids attended a great soccer camp a few years back at a local university and you recommended it. Perhaps you follow up with an email a month later to see how the camp decision panned out.

When we stay in touch on issues of interest, we show that we care. It shows we're thinking of them, and wanting to be helpful. Keeping in touch on issues is a useful way of staying in touch with people, and let them know we're thinking of them.

Remember Important Dates

We all like feeling special. Whether it's our birthday or work anniversary, our special milestones are important to us. Being remembered makes us feel good. It's become almost too easy to remember people's special days through Facebook and LinkedIn. We open up our computer and there it is right in front of us – today is Stephanie's birthday or Aaron's work anniversary. There is nothing wrong with being one of a few hundred people to "like" this, but how much effort does that really take?

Wouldn't it mean more if we received a handwritten note on our special day? How many cards do we actually receive these days, maybe ones from our mom, spouse, and kids? Nowadays, a personal note or card remembering someone's special day sends a much more powerful message. It shows that you care enough to remember, write, and mail a personal card. In an ironic way, actual mail has become more powerful than ever before.

Perhaps in the days before email and text messaging, an actual card wasn't that special. Frankly, it's been so long ago that we can't remember what it was like then. But today, taking the time to write a personal note demonstrates to a person that they're important to us and we're thinking about them. The next time an important colleague or client has a birthday or work anniversary announcement on social media, mark it in your calendar. Next year, try mailing them a card instead.

Give Small Gifts

Every year for Christmas I receive a gift in the mail from Cliff Farrah as a thank-you for serving on his company's board of directors. The gifts are branded with the Beacon Group logo. Beacon's branded gifts are really nice, not your typical corporate swag. And each one includes a personal note from Cliff.

Over the years I've received numerous quality items from Beacon, like an L.L. Bean canvas tote bag, Patagonia rain jacket, Cutter & Buck golf jacket, and many more. Each item, carefully chosen by the Beacon team, will last for many years and has useful value.

I wear or use Beacon's gifts nearly every day of the year. They do provide some branding benefit to Beacon, but the most significant benefit is that it shows that Cliff and Beacon care about me. And it serves as an opportunity to reach out and reconnect.

Gifts are an important tradition in every culture. The gifts don't have to be tied to a holiday; they can be offered on other important milestones as well. Perhaps we've just wrapped up a successful project together. How nice would it be to receive a small gift from a colleague? Maybe you know a client is taking a trip to Napa Valley for a wedding anniversary; perhaps you send a nice bottle of wine with a note congratulating them on their ten-year anniversary.

Here's my take on gift giving. The monetary value of the gift isn't important as long as the gift shows thought and care. In fact, in many governmental and business organizations' gift values are closely monitored with specific guidelines. Obviously, you'll have to abide by the policies in place with your organization and the client or colleagues.

Sending a gift of nominal value is an accepted business practice. The gift sends a message that you're thinking of the other person. And it's another natural way to keep in touch and stay top of mind.

Show Gratitude toward Others

Saying Thank You is one of the easiest and most underutilized ways we can demonstrate to others that we care. When someone goes out of their way to help or do something nice, showing that we appreciate their help or kindness closes the loop on the generosity. Learning how to gratefully receive the gifts of others is a way of showing that we care. I saved this idea for last – it holds significant personal interest to me.

I had no choice in learning the act of writing thank you notes; it was forced upon me early in life. By the time we could hold a crayon, my mom expected that my brother and I would write thank you notes. It didn't matter if it was a holiday gift, a birthday present, or a random act of kindness by a great aunt; you wrote a note by hand to say thanks. It's just what you did. It wasn't optional.

Naturally, as any kid would, I protested strongly. I would procrastinate for weeks in the hope that Mom would forget, but she never did. At some point, I gave in and took the five minutes it required to write the note. And Mom read each one before the envelope was sealed. If it wasn't up to her standards, the note was rewritten.

It wasn't until I was much older that I realized how important a thank you note could be. Perhaps it's because they are so rare that makes them stand out today. It's also one of the sincerest acts of empathy. Acknowledging a gift and displaying gratitude shows that we recognize the other person's kindness. It connects the giver and the receiver in a common bond of generosity and kindness, two important components in any relationship.

The act of giving thanks can take many forms. Maybe it's a handwritten note, or maybe it's a phone call. Depending on the nature of the thank you, perhaps a simple email is sufficient. Saying thank you allows us to stay in touch, while helping to deepen the relationship.

Offering genuine thanks applies to much more than receiving a gift. It is just as important when receiving a referral, an introduction, or some helpful advice. I was talking the other day with a professional friend on this topic. He explained that he had given a number of his co-workers referrals in the past year – referrals that he knew had led to new client work. Not once did his co-workers ever acknowledge the generosity with a call, email, or note. I could tell this hurt my friend's feelings. How hard would it have been to take a few minutes to say thank you? And imagine the benefits to the referral pathway in doing so.

Caring and Empathy Matter

All relationships – personal as well as professional – are based upon caring for others, showing genuine concern and empathy for another human being. When we remember others, it serves a number of important purposes in building trust-based relationships. It shows that:

- We're thinking of them.
- We care about them.
- We value their opinion and expertise.
- We want them to be successful.
- We're grateful for their help.
- The relationship is important to us.

These ideas are a just a few of the ways we can demonstrate that we care about others. You don't have to practice all of these – pick one or two approaches that feel right to you and stick with them. Make it a habit, like Herb Richardson's Sunday ritual. With time, our sincere interest and concern help build trust with those who matter most to us.

References

Wayne Baker quote: Adam Grant. *Give and Take: A Revolutionary Approach to Success.* Viking, 2013.

Adam Rifkin story: Adam Grant. *Give and Take: A Revolutionary Approach to Success.* Viking, 2013.

Herb Richardson story: Inspired by real characters known by the author.

Cliff Farrah story: Inspired by real characters known by the author.

Transparency Is Good, Right?

How and When to Be Transparent

Presence is about shedding whatever is blocking you from expressing who you are. The more we are able to be ourselves, the more we are able to be present. And that makes us more convincing.

—Amy Cuddy, Harvard psychologist and author, *Presence*

Transparency has received a lot of attention in business lately. Countless books on the importance of transparency have populated bookstore endcaps for the past decade. This trend began with *Transparency: How Leaders Create a Culture of Candor*, written by the late leadership guru Warren Bennis. Bennis, who served as a leading scholar at the University of Southern California for over 30 years and is widely regarded as one of the most respected pioneers in the field of leadership.

But what is transparency anyway? Transparency is a somewhat squishy business concept, like authenticity. While it's hard to nail down a definition, a good start for us is that transparency is synonymous with openness – the candid sharing of information with others.

It's generally accepted by most of us that more transparency is a good thing, especially in government and business organizations. Hiding information leads to a breakdown in trust between institutions and its stakeholders.

But when it comes to transparency in our professional lives, it can be hard to know how transparent we should be. It's kind of like knowing how honest to be when a friend asks, "Do you like my new haircut?" If transparency is good, is it always good? And, if not, when and where should we be transparent?

Pretending to be someone we're not requires too much emotional energy to sustain and we end up coming across as fake. Therefore, transparency is good when it allows us to share our authentic selves with others. Tying this to some earlier topics we discussed, transparency applies to our personal brand identity – what we want to be known for, our professional expertise, and who we wish to serve.

But our personal brands are much more than just our expertise – our brand identity is an amalgamation of everything about us. Being transparent also applies to personal qualities that make us human: our values, belief systems, interests, and hobbies. We learned that sharing these aspects of our personal lives helps us find common ground with others. And in finding common ground with others, we allow trust-based relationships to flourish.

Why Undersharing Is a Path to Irrelevancy

Earlier we heard from marketing guru Seth Godin on the importance of finding our *tribe*. We learned from Seth that humans are wired to seek out others who share common interests. This is why transparency is important, because it relates to sharing our personal interests and beliefs.

Some believe that it's a bad idea to be transparent about our personal interests in our professional lives. If we share with others who we are and what we believe in, we may turn some people off. And if we turn some people off by sharing who we are, then we shrink our pool of prospective clients. And if we shrink the pool of prospective clients, we have fewer opportunities for winning client business. Why would we want to be transparent?

I think this mindset of "never offend anyone" is a slippery slope to irrelevancy. I was recently listening to a podcast interview between Seth Godin and the editor of *Inc.* magazine. Seth said something, almost as a side comment, that really changed the way I think about transparency:

> Independent of whether you are an individual or a large company, you will either be judged or ignored.

When it finally sunk in, his comment helped me see more clearly why great brands take a point of view about things, such as Nike's support for Colin Kaepernick. I immediately grabbed my notepad and began sketching out a flow chart to illustrate Seth's point. (See Figure 20.1.)

When we're transparent about what we believe in and what our interests are, we are judged by others. This is scary for many of us; we were raised by caring parents who emphasized a philosophy of "not offending anyone." Don't stand out – fit in. Better to swim with the current. When it comes to business, though, not sticking out leads to being ignored.

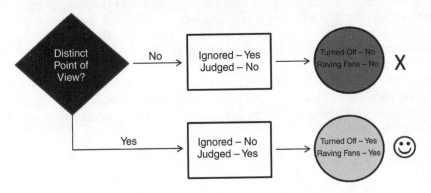

FIGURE 20.1 A Distinct Point of View Matters

When we are judged, people either love us or not. They're either drawn to what our brand stands for or not. And guess what? This is a good thing, because in turning people on we find our tribe of people who love what our brand stands for.

When we follow those who love our brand message – as shared through transparency – we find our tribe and begin to build authentic relationships with our audience. I call this *following the love*. When we are transparent with others about our interests and beliefs, some will love us; this is our tribe. Losing a portion of the market due to transparency is a much better alternative than being ignored by everyone.

When we aren't transparent about what our brand stands for – corporate or personal – we don't turn anyone off. It seems right, but it's not. By not being transparent about who we are and what our brand stands for, no one becomes a raving fan. So share your authentic brand and follow the love. Being ignored is a quick path to irrelevancy. Finding your tribe of raving fans is about being true to who you are. You may lose a portion of the market due to not having common ground, but you'll more than make up for the loss with a loyal group of colleagues who share your interests.

It's the reason you'll see a business put the Christian fish symbol in their store window. Chick-Fil-A – a privately held company – is closed on Sundays due to its Christian values. Those who are also Christians will be drawn in due to the common values. Those who aren't won't. Patagonia is another good example. Its founder, Yvon Chouinard, is an outspoken environmental steward. Those who align with Patagonia's strong environmental stance become loyal customers; others won't. It's better to be loved by 20% of the market than ignored by 100% of it.

I take a strong stand in my writing regarding certain business approaches that I believe in – for example, my views on cold-calling and likeability. I come out pretty strong in sharing my opinions and observations on these topics. Not everyone is going to agree with me, and I've had some interesting conversations with radio talk show hosts who disagreed with my point of view. It's better to find those who align with my belief system through transparency than having 100% of the market ignore me.

Being true to who you are is the purest form of transparency. You'll have to find the right way to be transparent that feels good to you. My personal belief is that it's best to be true to who you are, and share this openly with others, through small talk and in your personal bio, and you'll go a long way to forming trust with those you wish to serve.

How Much Is Too Much?

I draw the line, however, at using our business ecosystem as a dumping ground for our personal problems. If we become so transparent that every conversation ends up being a therapy session, then our professional colleagues will drift away from us. We all have problems, and most people tend to avoid those who overshare the difficulties of their personal lives. In this case, I believe being too transparent becomes an impediment to our success as rainmakers.

Clearly, my personal philosophy is not the only approach, but it feels right to me. I'll be open and honest about who I am as a person – what I believe in and what my interests are. But I won't openly burden every business colleague I speak with about the issues I'm dealing with in my personal life.

As with most squishy topics, there are exceptions. I follow my general rule most of the time with most people, but there are times when I do share personal challenges that I'm facing. This is strong medicine, so I use it sparingly and only with colleagues I know very well. When used appropriately with the right people, transparency on heavy personal issues can serve to strengthen relationships.

When I was going through a divorce about five years ago, my personal life was a mess. My divorce was as amicable as these events can be, but it was still a rough patch in my life. I had been married for 25 years and my identity was wrapped up in being a good family man.

Discussing my failed marriage wasn't easy for me. I was embarrassed, ashamed, and felt vulnerable. Most of us try to put a positive spin on issues in our personal lives. Some things, however, are impossible to sugarcoat, and I hid my situation from most people for many months. I did tell my closest colleagues. Over time, I shared my story with a few more with whom I had close relationships.

When the conditions were right, maybe over an after-work beer or a long dinner, I began to share my story with others close to me. Surprisingly, in many cases my transparency brought us closer together. In more than a few conversations, I discovered that others had also been through a divorce. Our common experiences strengthened our relationship.

Another example of transparency is with a close business colleague I've known for nearly 30 years. Over beers one night he revealed that his son had been expelled from college for selling illegal drugs. His son was currently in rehab, making progress and planning to enter a community college in the fall. His candid sharing of what his family was struggling through brought us closer. Through his transparency, he showed that he trusted me enough to talk about a difficult personal topic. For most of us, this is not a conversation we feel comfortable sharing with someone we don't know very well.

I keep my personal life issues out of my business relationships the majority of the time. I will share difficult personal challenges when appropriate with business colleagues I'm very close to, probably one or two dozen individuals. We all have our inner circle, those with whom we feel comfortable sharing our vulnerabilities. For the rest of my business ecosystem, I value their time too much to burden them with my personal issues.

I apply the same approach with regard to my feelings concerning a boss, a colleague, or an employer. We sometimes find ourselves working for a jerk, with a backstabbing co-worker, or for a firm that is highly dysfunctional. Transparency requires discretion in sharing with others these situations. Word travels fast, especially when it's juicy gossip.

If we're 100% transparent with everyone that our boss is a nut or that Sam is a sociopath or that our company is totally screwed up, we're just dumping baggage on others. And worse, it will come around to bite us when the word gets around. It always does.

Finding the Balance

This is my approach to transparency. Open with everything about my personal brand – what my passions are, who I wish to serve, my interests and beliefs. When it comes to my personal life problems and office gossip, I opt to be very selective with whom and when I share this information – it's strong medicine and best used carefully.

You'll have to find the right approach for you. This is not a topic that lends itself to a one-size-fits-all approach. I think with good judgment you will know who to be transparent with, when to be transparent with them, and what topics to be transparent about. When used appropriately, it is a powerful bond that can bring us closer together.

The Value of Saying "No" and "I Don't Know"

Transparency includes sometimes telling a client "No" or "I don't know." In a paradoxical way, telling a client these things often leads to greater trust.

When I recently caught up with a close friend, Susan, she shared a story with me about the importance of transparency when a client asks for something that isn't in the client's best interest. Susan told me it was not uncommon for her firm to walk away from a business opportunity if they weren't the best option for the job.

"We would tell the client straight up that we could do the work, but that we weren't the best choice," she said. "There were better or less expensive solutions. We offered to introduce them to other firms that would be a better fit for their specific needs. When we did this, it was interesting to see the client relationship become stronger. In an odd way, in being completely candid and declining the offer to help, we became a more trusted advisor."

Eugene Buff – the innovation consultant in Boston we spoke with previously – shared that his approach is to first understand what the client needs. He said, "If I can help them, great. If not, I am honest with them. I tell them when I don't have the experience or expertise to help them solve their problems. I don't want to fail my clients. My reputation stands on my ability to help my clients. If I accept a project where I can't succeed, it's going to lead to an unhappy client and a bad reputation for me. Better to tell them the truth and maintain trust."

This type of transparency is at the core of professionalism. We're not serving anyone – our clients or ourselves – if we tell the client yes when we really should say no. Charles Moren, the R.L. Matus sales engineer we met earlier, shared a related point regarding transparency in saying "I don't know." "When a client asks me a question and I don't know the answer, I tell them the truth. I'll say, "I don't know, but I'll find out the answer for you." I make a point to get back to them by the end of the day. I'll research the question and provide them with an answer."

It's so tempting to take on the new client work. Clearly it would be in our short-term financial best interest to do so. It's also hard for us to say, "I don't know," especially when we're supposed to be the expert. Ironically, saying "No" and "I don't know" builds trust. In these two areas, transparency is always the best approach.

References

Seth Godin quote: "Real Talk: Business Reboot." *Inc.*, interview published May 20, 2020.

Chick-Fil-A story: "The Church of Chicken: The Inside Story of How Chick-Fil-A Used Christian Values and a 'Clone Army' to Build a Booming Business That's Defying the Retail Apocalypse and Taking Over America." *Business Insider*, August 8, 2019.

Yvon Chouinard story: "The Patagonia Adventure: Yvon Chouinard's Stubborn Desire to Redefine Business: An Inspiring Business Leader Who Built a Clothing Brand Dedicated to Environmental Stewardship." *B The Change*, September 6, 2016.

Susan story: Inspired by real characters known by the author. Names and locations have been changed.

Charles Moren quote: Interview by *How to Win Client Business* research team, 2020.

Skill 5: Practice Everyday Success Habits

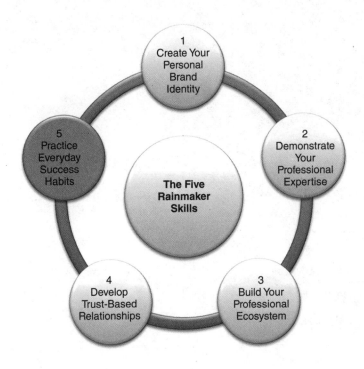

CHAPTER

21

The Daily Habits of Successful Rainmakers

The One Hour Each Day That Will Build Your Career

A successful life does not result from chance nor is it determined by fate or good fortune, but from a succession of successful days.
—Ari Kiev, MD, author, *A Strategy for Daily Living*

We've covered a lot of ground since beginning this quest together. The five rainmaker skills include a broad range of interconnected fields including human psychology, consumer behavior, marketing strategy, and behavioral economics, each illustrated through stories from those who have walked this path before us. We have weaved into our journey a bit of history and philosophy to bring color to our landscape.

Each of the skills we've covered thus far is important to our success in becoming rainmakers:

- Skill One: Create Your Personal Brand Identity
- Skill Two: Demonstrate Your Professional Expertise
- Skill Three: Build Your Professional Ecosystem
- Skill Four: Develop Trust-Based Relationships

The fifth and final skill, *practice everyday success habits*, may be the most important. Without applying these skills in our daily lives, it's just theory – nice to know, for sure, but having little impact on our careers. The first four skills are about *what to do*; the final skill is about actually *doing it*. This is what sets the rainmakers apart from the rest of the pack.

This reminds me of a quote I read early in my career, in a 1989 issue of *Harvard Business Review*:

> Ideas alone are not good enough. Ideas are not deeds. Ideas are rarely converted into action unless proselytized with zeal, carried with passion, sustained by conviction, and fortified by faith. They need authentic champions. Above all, ideas need people who are doers, not just talkers.

Practicing the rainmaker skills is hard because they are not the most urgent priorities of our day. Daily life sets in and we get busy doing our work. Urgent tasks squeeze out the important.

The other complicating factor is that most of us want to practice our craft, not market ourselves. We've studied and worked hard for years to become successful accountants, architects, financial planners, and organization designers. Our natural default is to focus on the work at hand.

Most of us enjoy our work; it challenges us and we get to solve important world problems. And we get paid pretty well in doing so. But the hard reality is that in order to make partner or succeed in our own firms, we have to learn how to win client business. If we don't, our careers will stall out mid-career – we'll get pushed to the curb or into a dusty back-office cubicle.

Most of us were never taught how to win client business. We received no formal training from our universities in how to become a rainmaker. Nor did we receive much effective mentoring, because successful partners have a hard time in explaining to others what they do and why they do it. And when rainmakers do offer us advice, what worked for them may not be the best approach for us.

To help us learn how to win client business, we've had to overcome several preexisting mindsets. We had to unlearn preconceived thoughts on what a rainmaker is in order to become one.

Mindset Shift One: Seeing ourselves as problem-solvers instead of salespeople

Mindset Shift Two: Viewing our professional accomplishments as credibility markers rather than bragging

Mindset Shift Three: Thinking of our professional network as authentic relationships rather than names in a database

Mindset Shift Four: Using our professional ecosystem to make new friends is highly appropriate when it's based upon a genuine desire to help others

Armed with these new skills and mindsets, we now arrive at the hardest part: putting these together into practice.

The Roller Coaster of Revenue

There is value in practicing these rainmaker skills – even if performed irregularly. If we write a timeless book, for example, it may have lasting value even if the practice of writing is dropped. Look at David Maister's *The Trusted Advisor* – this book is 20 years old and remains a best-seller to this day. A good presentation at a conference a year ago may still generate a few clients today. Or a single relationship you built decades ago continues to feed you and your team. Perhaps a charity board you were once on continues to open doors for you today.

Here's the trouble, though. The vast majority of our rainmaker activities have a shelf life – whether it is a paper, speech, a board role, or personal relationships – and without consistent stewardship their benefits will fade with time. And when this happens, the gates on the client pathways begin to close and revenue tapers off.

This is the life that many professionals live. We crank up the rainmaker machine, generate a project or two, get busy doing good client work, and the machine goes in the closet. Six months later everyone in the office begins to get nervous as the work dries up. Partners begin to panic – dust off the rainmaker machine again and rush out to find the next client.

Not only is this a stressful way to work, it's also disingenuous. When we call upon colleagues only when we need something, they will eventually stop returning our calls. We demonstrate that our relationships aren't that important to us – except when we need them to be. This is not the behavior that signals to others that we really care about them. Relationships are built upon empathy, care, and honesty, and these are not characteristics that lend themselves to sporadic attention.

So what's the right approach to consistently winning client business? The approach of successful rainmakers is to practice these skills every day of the year. It becomes a natural part of the daily routine. Regardless of how much client work there is to do, the most successful among us commit to practicing these skills consistently. It's the only way to build a steady pipeline of work. Otherwise, we live on the revenue roller coaster.

Practicing the Rainmaker Skills Every Day

Here's some good news for you if you're early in your career: the best time to begin learning the rainmaker skills is years before you are expected to generate revenue. If we begin practicing these skills before we're expected to demonstrate them, we'll develop the skills gradually. We'll find the approaches that work best for us. The longer we wait to build these into our daily routines, the harder the transition will be. And it's certainly harder to learn these skills when the pressure to deliver revenue is upon us.

The best rainmakers practice these skills every day. Others cite lack of time as an excuse. Ford Harding speaks to this in his excellent book *Rain Making*:

> Most professionals failed to complete the marketing tasks they promised, always with the same excuse – they were too busy. By any fair measure, they were busy. But so were the few who marketed and met the objectives they had set. It was always the same few, and they were no less busy than those who had made excuses. Somehow they had found a way. The executive committee noticed this, and rewards went to those who marketed.

When we observe the success of others, we often chalk it up to extraordinary talent or luck. Whether actor, writer, athlete, or rainmaker, the reality is that behind their success is a daily habit of practice. Their success is not the result of a singular moment of inspiration, but the output of many years of focused effort.

We don't often see the hours of committed effort in our rainmakers; what we see are their success and recognition. When asked about his Sistine Chapel masterpiece, Michelangelo responded, "If you knew how much work went into it, you would not call it genius." So begin the habit of practicing the rainmaker skills every day. It's the only approach that I'm confident will build a strong, lasting career and good reputation.

The Most Important Hour of Your Day

Laura Vanderkam has made a career out of studying the daily habits of successful people. She shares these findings in her engaging book *What Most Successful People Do Before Breakfast*. She observed a pattern in the ways in which many successful people start their day. They begin with an hour of reflection and planning before phones start ringing, emails start whirling, and countless other distractions fall upon us.

For some, this is a morning walk with the dog or a jog through the park. For others, it is with a cup of coffee and a journal. Some meditate or pray. Independent of the habit, the quiet ritual serves the same purpose: it allows us to begin each day with purpose – identifying the most important activities for the day.

I've heard similar stories from countless professionals who are at the top of their field. Here's how the CEO of a successful strategy consulting firm described his morning routine:

> I begin each morning with an hour of time alone with a clean sheet of paper on my legal pad. At the top, I write: The Most Important Things to Do Today. I have four life categories: health, service, wealth, and enjoyment.
>
> I list the things I am going to do for my health – eating well, hydrating, exercising, sleeping, etc. Next, I list the ways I am going to serve others – to contribute to the success of my team, clients, colleagues, etc. I then outline my intentions for building my financial future – earning, saving, and investing. Lastly, I include time for the things that bring balance to my life: dinner with family/friends, reading, playing tennis, or attending my kid's soccer game.
>
> I keep this legal pad with me all day. In between meetings or calls, I take a few minutes to revisit it. Things don't usually work out exactly as I hoped, but I've found that it helps me stay focused on today's most important things.

Left to chance, urgent things always crowd out the most important. This simple life fact has created a whole industry out of self-help and time management advice. Countless books, seminars, and programs are available to help us get more out of our day. Fortunately for us, there is a lot of valuable advice from the experts that we can apply to our daily lives as developing rainmakers.

The Franklin Time Management system was the first training program I attended upon joining GE's leadership development program. Those of you who were around in the 1980s probably remember this program, and may have attended one. The Franklin program was named after Benjamin Franklin, who kept a small private book with his daily thoughts, intentions, and observations. A core technique of the Franklin approach involved beginning each day with 15 minutes of "solitude and planning."

You may notice a connection between the Franklin program that I attended over 30 years ago and the wisdom in Laura Vanderkam's book: many of the most successful people start their day with quiet time that allows them to begin their day with a plan for success.

If we are to build the rainmaker habits, we have to start with a daily plan. With time, this will become as normal as tying our shoes before heading out the door in the morning. If we don't start with a plan of the day's most important tasks, they're unlikely to get done. And, as my CEO friend found, every day won't go exactly according to plan, but our days will trend in a positive direction by starting each day with thoughtful intent.

References

Ford Harding quote: Ford Harding. *Rain Making: Attract New Clients No Matter What Your Field.* Adams Business, 2008.

The Most Important Hour of Your Day story: Laura Vanderkam. *What the Most Successful Do Before Breakfast.* Portfolio/Penguin, 2012.

CEO's morning ritual story: Inspired by real characters known by the author. Names and locations have been changed.

CHAPTER

22

Making the Rainmaker Skills Stick

66 Days That Will Shape Your Future

Each day must be approached as a unit; each must be lived with care; and if this is done, the procession of days will turn out all right.
— Louis L'Amour, writer, recipient of the Presidential Medal of Freedom

Dr. Maxwell Maltz noticed an interesting pattern among his surgery patients in the 1950s. What he found was that when he performed an operation, it took about 21 days for the patient to get used to the changes. When a patient had a leg amputated, for example, Dr. Maltz found that the person would sense a phantom limb for about three weeks before adjusting to life without it. According to Dr. Maltz, "These phenomena tend to show that it requires a minimum of 21 days for an old mental image to dissolve and a new one to form."

And that's how the "it takes 21 days to form a new habit" myth began. You see this in the self-help advice from gurus like Zig Ziglar and Tony Robbins. But new research suggests that it takes much longer.

In 2009, Dr. Phillippa Lally and her team at University College London set out to determine just how long it takes to form a new habit. In a study of 96 people over 12 weeks, Lally's team had each individual practice a new habit and report daily on whether they performed the new behavior and if it felt automatic.

Some people chose simple habits like "drinking a bottle of water with lunch"; others chose more difficult tasks like "running 15 minutes before dinner each evening." Here's what the researchers found: on average, it takes more than two months before a new behavior becomes automatic – 66 days to be exact.

Learning from the latest research, it will likely take us longer than 21 days as once believed to form the new rainmaker habits. If we commit to a three-month period – figuring the average month has roughly 22 working days – we'll build into our lives the daily habits demonstrated by successful rainmakers.

Building the Rainmaker Skills into Our Daily Lives

When we start to build the rainmaker skills into our daily lives, we'll discover that some skills deserve daily attention and others lend themselves to weekly or monthly reflection.

Create Your Personal Brand Identity

The essence of your brand identity is what you want to be known for and who you wish to serve. Naturally, this requires a lot of thought early on as you settle on which niche you want to become the go-to expert. We learned that we must be in the top three when considered for new client work, or we won't get the business. And we discovered the best way to achieve this is by "shrinking the pond" until we become a big fish. Over time, as we become more successful, we can grow our pond by expanding into adjacent market niches.

Your professional focus may evolve over time as your interests and market conditions change, but your chosen expertise won't change daily. Early on in the process of learning these new rainmaker habits, there is aspirational value in reviewing your brand identity each morning. In this practice, you'll revisit what you want to be known for and whom you wish to serve. Write these down and repeat them every morning for 66 days.

1. I want to be seen by my clients and colleagues as the go-to expert for [*insert expertise*].
2. My aim is to be among the very best in serving [*insert target audience*].

There are vast quantities of research to support the positive benefits of mentally rehearsing successful outcomes. It's a big part of the psychology of coaching elite performers in every profession. We have to believe we can do something before we can achieve it.

Dabo Swinney, head coach of the Clemson University football team, says, "If you want to be an overachiever, you have to be an overbeliever. It's amazing what your team will accomplish when you believe in them." Dabo may coach football, but he also recognizes he's a sports psychologist. He understands that achieving greatness begins with believing in ourselves.

A big part of our success occurs between our ears. Mentally rehearsing the brand identity we aspire to become is effective in helping us achieve the success we desire. It will help reinforce what brand identity we are pursuing. As we progress, reviewing our personal brand weekly or monthly will help us see when it needs changing. If we find that our personal brand identity no longer aligns with what we believe and how we view ourselves, we'll know it's time to adjust our goals and aspirations.

Demonstrate Your Professional Expertise

Clients grasp for clues to see that we are genuinely good at what we do. And they frequently struggle to find the expert whose approach, beliefs, and philosophies fits well with theirs. Clients seek to avoid regret, and search for signs that we are good at what we do. We found that it's not that our sales cycles are long, but rather that it's a long buying cycle for many clients.

We examined eight proven techniques for demonstrating our expertise to prospective clients:

- Technique 1: Writing
- Technique 2: Public speaking
- Technique 3: University teaching
- Technique 4: Radio programs and podcasts
- Technique 5: Serving on a board of directors
- Technique 6: High-profile work and case studies
- Technique 7: Industry awards
- Technique 8: Professional certifications

Clearly, these things won't happen in a day. And they won't happen without making them happen. No one becomes a rainmaker by accident. So spend time each morning in your quiet time investigating the approaches that feel right to you. Pick one or two approaches, and begin working to make them happen.

Writing a blog, joining a board, or teaching a university class won't happen today. But they can begin today. Seek help from others who have had success in using the approaches you are interested in. And remember, it's much better to pick one or two approaches at a time and be consistent with them than to dabble a little in everything.

Build Your Professional Ecosystem

The lone wolf in professional services is rare. Successful rainmakers build an ecosystem of people who work together to support one another. We work as a team of committed people and together we can achieve great things. The preferred alternative to cold-calling is opening up the referral and inquiry pathways – the source of the majority of new client business. And these pathways are opened when we build an ecosystem of others we believe in and who believe in us.

Rainmakers make friends in natural ways, the way we once made friends as kids. Rainmakers find the activities and organizations they are interested in and get involved. When they can't find the right group or club, they form a new one. And, in getting involved, we rub elbows with others who share common interests and beliefs. Your ecosystem will start with:

- Clients, current and former
- Colleagues in your company or firm
- Business partners/alliances
- Colleagues in related professions
- Industry association members
- Professional organization members
- Community officials
- Former classmates
- Friends and neighbors

Begin your day by brainstorming the ways you can meet people organically. What are your interests and passions? What professional organizations resonate with you? What community organizations align with your values and beliefs? Those most successful at building their ecosystem are authentic in their involvement in these pursuits. They join because they believe in the organization's purpose and mission, and along the way meet others who share these beliefs.

In your morning's quiet time, you'll find clarity on what is most important to you. Look for others you know who are involved and ask for advice. Not sure what organizations or activities to pursue? Explore the various options. With time you'll find the things that feel right to you. And in finding what interests you, you'll find your tribe. You'll begin to make friends with people who you have much in common, and begin to build your ecosystem.

Develop Trust-Based Relationships

Clients hire people who they come to know, respect, and trust, or who come highly recommended by a good friend or colleague. Respect comes from our professional credibility. Trust comes from demonstrating to others that we care, that we are interested in their success and have their best interests at heart.

Relationships begin to take root when we get to know people personally. This process is enhanced with small talk – the seemingly irrelevant chit-chat that occurs at the beginning of many business conversations. Finding common ground is a natural way of learning about other's interests and sharing with others ours as well.

When we find common ground with others, relationships begin to develop. We learned that listening is of one of the purest expressions of empathy, and a skill that can be developed by asking good questions and being curious about other's interests.

Keeping in touch shows that we care. And it also serves the important role of keeping us top of mind, opening up the pathways of repeat business, referrals, and inquiries. We can stay in touch with those in our ecosystem in many thoughtful ways:

- Recognize other's success
- Provide helpful news/information
- Offer introductions and referrals
- Ask for advice or guidance
- Follow up to inquire about an issue
- Remember important dates
- Give small gifts
- Show gratitude toward others

Developing trust-based relationships isn't hard. But we discovered that it isn't common, either. It begins when we care about others first, and put their interests ahead of our own. When we show sincere care for others and seek to help support their success, we build trust. And when we build trust first, we receive the benefits of building mutually supportive relationships.

Begin each morning by thinking about others you can help. Start each day with a list of three to five things you can do today. Plan these small gestures into your day. Maybe we call a business partner to say thank you. Or invite a colleague to lunch to ask for advice. Or send a friend a news clipping or a good book recommendation.

The list of ways in which we can demonstrate to others that we care is long. But if we don't plan these into our day, they won't happen; they'll be crowded out by a hundred other urgent to-dos. Practiced daily, over time, we'll build relationships based upon trust and respect. And our thoughtfulness will come back to us in the form of care and kindness from others.

Practice Everyday Success Habits

Unless you're an avid runner, you've probably never heard of Jim Ryun. He's not famous – at least outside of Kansas or the running community – but he demonstrates the kind of persistence necessary for achieving success. ESPN named him the best high school athlete of all time, beating out stars like Serena Williams and LeBron James. Wait, what?

Jim became the first high school athlete to run a mile in less than four minutes – a feat many doctors and coaches did not think was humanly possible in the mid-twentieth century. At his peak, Jim was considered the world's top middle-distance runner – the track events longer than a sprint but shorter than a 10K. Jim is the last American to hold the world record in the mile.

Born in Wichita, Kansas, in 1947, Jim didn't demonstrate any early talent in sports. When asked how he found running, he said, "I couldn't do anything else. I was cut from the church baseball team, the junior high basketball team, and I couldn't make the junior high track and field team. I found myself trying out for the cross-country team and running two miles even though I'd never run that distance before. To my surprise, I made the team…and, that's how it all began."

What were the keys to Jim Ryun's running success? He had natural talent, for sure. Maybe undiscovered at first, but it's hard to run a four-minute mile or win an Olympic medal without natural abilities. Yet Jim's coaches consider his persistent, daily approach to practice as one of the keys to his many running achievements.

When asked about his disciplined approach to running, this is what Jim had to say:

Motivation is what gets you started; habit is what keeps you going.

Habits are powerful things. When we build them into our daily routines, we can achieve things that seem incredible. All of us have natural talents for our craft or we wouldn't have made it this far in our careers. If you don't have a mind for numbers and attention to detail, you won't make it through accounting school and pass the CPA exam. If you don't have some natural design abilities, you won't make it through a master's program in architecture. The same goes for engineers, attorneys, and the other professions as well.

But like Jim Ryun, our natural talents will only take us so far. We have to build daily habits into our lives to achieve success. Habits can be constructive or destructive. Good habits can help us reach our aspirations and bad habits can prevent us from achieving our potential.

And this wisdom is not new. Ovid was a Roman poet during the reign of Augustus in the first century BCE. He wrote, "Nothing is stronger than habit." And before Ovid, the Greek philosopher Aristotle, in the fourth century BCE: "We are what we repeatedly do. Excellence, then, is not an act, but a habit." And before ancient Rome and Greece, the power of habit found its way into early Asian philosophy. In the words of Lao Tzu, an ancient Chinese philosopher:

> Watch your thoughts, they become your words; watch your words, they become your actions; watch your actions, they become your habits; watch your habits, they become your character; watch your character, it becomes your destiny.

The habit of starting each day with an hour of quiet planning time has proven to be a key in the lives of many successful individuals before us. And new behaviors take 66 days on average to become habit. Great things can happen when we establish habits that keep us going, long after the initial motivation has worn off.

Get started today in building the rainmaker skills. Begin with one hour each morning. Practice these every day for three months and you'll begin to form strong habits. And these habits will assist you in building a successful career doing what you love with those you wish to serve.

References

Dr. Maxwell Maltz story: James Clear. "How Long Does It Actually Take to Form a New Habit? (Backed by Science)." www.jamesclear.com.

Jim Ryun story: Mario Fraioli "The Best Ever: Exclusive Interview with Jim Ryun." *Competitor*, March 17, 2014.

Ovid quote: Nathan Brooks. *The Metamorphoses of Publius Ovidius Naso*. A.S. Barnes & Co., 1857.

Aristotle quote: J.L. Ackrill. *Essays on Plato and Aristotle*. Oxford University Press, 1997.

Lao Tzu quote: Roel Sterckx. *Ways of Heaven: An Introduction to Chinese Thought*. Basic Books, 2019.

CHAPTER

23

Finding Your Rainmaker M.O.

Building a Rainmaker System That Works for You

Everyone should carefully observe which way his heart draws him, and then choose that way with all his strength.

—Hasidic saying

Timothy is a senior partner at a large accounting firm in Washington, D.C. He and his wife are empty nesters – their three kids are all out of college and pursuing careers. Timothy wakes up each morning about 7:00 a.m., fixes a cup of coffee, and plans out his day. He's usually in the office by 9 after a short walk from his Alexandria townhome.

Timothy has a small stack of index cards with the names of each of his closest business relationships – his Tier 1. He has a second, larger stack for his secondary relationships – his Tier 2. He carries these cards with him everywhere. He may even sleep with them. (I hope he has the information backed up somewhere.)

Each morning Timothy pulls three cards off the top of his first stack. He makes a point of reaching out to each of these people during the day. He may invite one to lunch, call another, and write a thank you note to the third. But somewhere in the day – maybe in the minutes between client meetings – he always finds time to contact each person.

At the end the day, Timothy makes a note or two on each of today's cards and moves them to the bottom of the stack. The next day, he begins with three new cards from the top. By the end of the month, he's connected with each of his closest contacts in some meaningful way. He does the same with his other stack as well. The second stack of index cards is larger, and the length of time in cycling through this group is longer – about 3 months.

Timothy has found a system that works for him. While incredibly low-tech, his index cards are highly effective at keeping in touch with his professional eco-system and remaining top of mind. It works because he is consistent with it, and it fits his natural strengths and preferences.

The Modus Operandi

Mrs. Mary Clara Blackmon walked to the chalkboard at the front of the classroom. Across the top she wrote the words *Modus Operandi*.

Our *modus operandi*, class, is our method of operations. Each day we will follow our *modus operandi* to give each day structure. We will begin with roll call. After that, we'll have a short lecture on the topic of the day, followed by the day's in-class assignment. When you've finished your assignment, raise your hand and I will come to your desk and grade it. Once you've success-fully completed today's assignment, you may work on the homework for tomorrow.

Mrs. Blackmon was my eighth-grade science teacher. She was eccentric in the best sort of way. She wore wild dresses. Her face was framed by large, black artist's glasses, the kind you might see on a fashion designer. Her knowledge of the world seemed endless – she layered Swahili into her lessons on geology and Nietzsche into her discourse on the solar system.

Her love of all things intriguing sparked our curiosities. I remember being struck by how much I didn't know – and wanted to learn. She was strict and demanding. She also loved each of us and wanted us to succeed. We strived to rise to her expectations.

Each rainmaker has a unique modus operandi, or M.O. No two are alike. Timothy found his rainmaker M.O. You, too, will find yours. Timothy's M.O. was a simple index card system. He followed his system as faithfully as we followed Mrs. Blackmon's. And your M.O. will bring structure to your day – a routine that can be as faithfully followed as lather, rinse, and repeat.

A More Modern M.O.

Amy is the co-founder of a mid-sized PR firm in Madison, Wisconsin. Her M.O. is considerably different from Timothy's, but equally effective. Amy is tech-savvy, although she won't admit it. She cannot live without her MacBook, iPad, and iPhone – they are rarely out of arm's reach.

Amy wakes up around 6:00 a.m. and takes her two dogs for a walk by the lake. She comes back, has a cup of coffee with her husband, and helps get her son ready for school. She's typically in the office by 8:30, and spends the first 30 minutes there quietly planning her day.

Starting at 9:00 a.m., her mornings are filled with client presentations, staff meetings, and assisting her team in writing proposals. By 3:00 p.m., she's done with the heavy lifting for the day and schedules time in her office for her daily client development activities.

Amy relies on a cloud-based CRM (customer relationship management) platform for helping prioritize her daily client outreach. Her system organizes her contacts based upon how often she wishes to reach out to each. The system allows her to keep track of important dates and other personal information for each person in her network.

The CRM system creates a daily activity report of the people she needs to reach out to. It also keeps track of the last conversation and key topics to follow up on. Amy is committed to each day's outreach activities. She typically does these in her afternoon quiet time, unless the person she's connecting with prefers mornings. Then, she'll work these in between her other meetings before lunchtime.

While Amy's M.O. is quite different than Timothy's, her system allows her to keep in touch and stay relevant with those in her professional ecosystem.

The Middle Ground

Stephan is a senior partner in a large Dallas law firm. Stephan likes to get up when things are quiet in his home – typically around 5:00 a.m. His M.O. begins with an hour of coffee, reading the news, and planning his day. He'll then go for a jog and have breakfast. He's typically in the office by 8:00 a.m.

He finds his most creative and productive hours to be earlier in the day. He blocks out four hours in the morning to focus on the things that require his best mental effort. Stephan saves the afternoons for phone calls, meetings, and keeping in touch with the people in his ecosystem.

His rainmaker system falls somewhere between Amy's and Timothy's. His is more tech-savvy than Timothy's index cards but definitely less sophisticated than Amy's cloud-based CRM system. He uses a spiral-bound notebook for his morning planning; he uses his computer for reminders, scheduling, and contact database.

Each Sunday evening he lists the people he needs to reach out to in the coming week, and he enters these into his desktop calendar on Monday morning. His approach of who he needs to speak with isn't as methodical as Timothy's or Amy's. His list is more gut feel. Maybe it's a call to a colleague, or an introduction for a friend, or a thank you note. Maybe it's sending a small gift to a client.

Stephan has found a rainmaker M.O. that works for him. His approach keeps him in touch with those he cares about and wishes to serve.

Finding Your M.O.

My good friend Dan is the CEO of a successful transportation company, the largest charter bus company in our region. Dan is a morning person. He's up by 5:00 a.m. and it's all systems go. He's a flurry of activity from the moment his rises. He's grabbing a smoothie to go and in the gym by 5:30. By 7:00 a.m. he's in the office making things happen.

My day unfolds much differently. My body and mind wake up slowly. I may be in the office by 8:00, but it's been a gradual, unhurried process. Dan's day begins like a shot from a cannon; mine is like a caterpillar crawling across the lawn.

For many years, I tried to force myself into a routine that looked like Dan's. I thought this was the way successful people operated. It took me many years – more than I'd like to admit – before I realized there isn't one right way for a successful person to go about their day. The key is to find a rhythm and routine that plays to our natural strengths.

Your M.O. is your own unique way of practicing the rainmaker skills – influenced by your personality, style, and daily routines. All successful rainmakers have a system for their client development activities. In the madness that often unfolds on any given day, there is always an underlying *modus operandi*. There are an endless variety of approaches; the best one is the one that works for you and that you'll stick with.

Rainmaker M.O. Best Practices

I've seen highly successful rainmakers who rise early with the rooster and others who work into the wee hours like a nightclub bouncer. I've seen some who never use a computer and others who swear the key to their success is a sophisticated software system. You'll find an M.O. that works best for you – perhaps with a bit of trial and error. While there is no one right way, we can apply some of the best practices of the rainmakers who have come before us.

- Develop a rhythm to your day that fits your natural preferences.
- Commit to beginning each day with time for reflection to set your intentions.
- Develop a daily routine – high-tech or low-tech – for building your professional ecosystem and staying in touch with the people who matter most.
- Set aside time each week to reflect upon your personal brand identity and how you demonstrate your professional expertise.

In sticking with your M.O. for three months, you'll establish new rainmaker habits. And when something isn't working, you'll modify your approach, knowing that enlightened trial and error is the best teacher in finding a system that is right for you.

References

Timothy story: Inspired by real characters known by the author. Names and locations have been changed.

Mary Clara Blackmon story: Inspired by real characters known by the author.

Amy story: Inspired by real characters known by the author. Names and locations have been changed.

Stephan story: Inspired by real characters known by the author. Names and locations have been changed.

Dan story: Inspired by real characters known by the author.

The Rainmaker's Journey

Thoughts on Becoming a Rainmaker

Stop Trying to Be Wonder Woman or Superman

Don't ask what the world needs. Ask what makes you come alive. And go do it. Because what the world needs are people who have come alive.
—Howard Thurman, American civil rights leader

We've all had heroes in our lives. The women and men we've looked up to, aspired to become, and wanted to be like. A funny thing happens, though, when we get to know our heroes more closely: at some point we realize they are humans just like us. In addition to their many admirable qualities, they also have bad habits, quirks, and weaknesses. When we take our heroes down off of their thrones, we stop thinking of them as heroes at all.

Rainmakers often take on a mythical persona in the office. Part of their mystique comes from us believing they are magic, wizards in tinseled capes. They work in mahogany-paneled offices, drive S-Class Mercedes, dine clients at the finest restaurants, and fly first class. Meanwhile, we're working in a tiny office, driving a

Toyota, eating Chinese take-out, and sitting at the back of the plane. It's no wonder we're a bit envious.

But here's the thing. Rainmakers are not wizards. And they aren't Wonder Woman or Superman. Rainmakers are real people just like you and me. Behind the lore and legend of the rainmakers are people who once worked late into the night dining on leftover pizza and driving home in rusty hatchbacks.

What We Can Learn from the Beatles, Led Zeppelin, and the Rolling Stones

I teach business strategy to college seniors at my local university. It's a topic I've been drawn to since my mid-20s. Along the way I've studied hundreds of great companies, read countless books, and attended lectures by the best scholars. And at the end of all of this, I have arrived at a belief that striving to be the best is a failed strategy for you, me, and every business out there.

What? Isn't success about being the best? Contrary to popular belief, strategy is not about being the best. Being the best in the world is akin to being Wonder Woman or Superman. Sounds good, but's it all myth. There is no best. What does it mean to be the best anyway? The best at what? The best how? The best for whom?

The way we got to this flawed mindset was in seeing business as war. For thousands of years historians have studied the strategies of successful generals and armies. And for good reason: the outcome of wars shaped the world in which we live: the boundaries of our nations, the languages we speak, and the cultures of our people. When it came time to think about business strategy, scholars went to what they knew best – war strategy. But business is not war.

Harvard economist Michael Porter has been the preeminent scholar on business strategy for the past three decades. Strategy is the focus of his research and he has written many excellent seminal books on the topic. Porter's books are dense and heavy, and include more than a few mathematical equations.

You can read Porter's book if you're so inclined, but another approach is to read Joan Magretta's synopsis of his theories in *Understanding Michael Porter: The Essential Guide to Competition and Strategy*. Porter's ideas are all here, but they're presented in a manner that is much more accessible.

Magretta shares this Porter insight:

> **In war there can only be one winner. Victory requires that the enemy be crippled or destroyed. In business, however, you can win without annihilating your rivals. In business there can be multiple winners.**

Walmart has been a leader for many years in discount retailing, but so has Costco. Delta has been a successful airline, as has Southwest. Budweiser is the number one beer brand in America, but Sierra Nevada is the leading family-owned microbrewer. There is no best in business. There are just great companies doing what they do best for the people they aim to please.

Magretta goes on to explain that some strategists have compared business to sports. She explains Porter's belief that "The sports analogy is just as misleading. Athletes vie with each other to see who will be crowned 'the best.' They compete to win. But in sports, there is one contest with one set of rules. There can be only one winner. Business competition is much more complex, more open ended and multidimensional."

Businesses don't play by the same rules as athletes do in sports. Does Amazon play by the same rules as Barnes & Noble? Does Airbnb play by the same as Marriott? Does Tesla use the same playbook as GM? Furthermore, sporting events have an end. In business, there is no end in sight – just another week, month, quarter, and year.

According to Porter, "A better analogy than war or sports might be the performing arts. There can many good singers or actors – each outstanding and successful in a distinctive way." This insight forever changed the way I think about business strategy.

I apply this insight in my first university class each semester. I ask my students which band is better: the Beatles, Led Zeppelin, or the Rolling Stones? This always creates a lively debate. Then, as the debate begins to wind down, a student sitting quietly in the back asks, "What do you mean by best?" I've been waiting patiently for this moment. I smile – as good teachers do – and say, "Precisely!"

There is no way to say which band is better unless we ask, "Best how?" And "Best for whom?"

Strategy isn't about trying to be the best. Strategy is about figuring out what you want to be really good at, and who you want to serve. And, just like performers in the various arts, we get to define how we want to be best, what we want to be best at doing, and who we want to be the best for.

Becoming a Rainmaker Is Not a Structured 10-Step Program

I don't offer a cookie-cutter approach to winning client business because I don't believe in one. I don't believe in a formulaic approach that tries to make you become like someone else: watch me, do this, and you'll be successful, too. A 10-step program will work for some – those who happen to fit the right profile the program was designed for. For many, the programmed approach won't work because it isn't suited to their strengths and preferences.

I've observed successful rainmakers from every imaginable service. And what I've found is that no two are alike. If you work for a larger firm, you'll have many partners who have come before you. And if you watch what they do, you'll see they each do it differently. You'll discover there is no one right way. There is no best. There is just what is best for you.

Just as there are countless great bands each creating the music they love and building their own fan base, each one of you will become a rainmaker in your own way, doing what you are best at and serving the people you want to serve. You'll put your own twist on things, infuse it with your unique personality and point of view, and along the way you'll find your tribe – the people who get you, see what you're about, and understand what you aspire to become.

The rainmaker approach I believe in is tied to an understanding of how clients buy. It begins with empathy for the prospective clients who have to make difficult choices about who they wish to hire. Clients want to make good decisions; it's just not as easy as buying a home or a car.

The approach that I believe in is tied to the client's buying decision journey. Clients need to:

- Be aware of you
- Understand what you do and who you serve
- Have an interest in your services
- Respect your professional abilities
- Trust that you'll have their best interests at heart
- Have the funds and authority to hire you
- Are ready to begin because the timing is right

And, as guides in their journey, we will practice the rainmaker skills in our own way. We will:

- Create a personal brand identity
- Demonstrate our professional expertise
- Build our professional ecosystem
- Develop trust-based relationships
- Practice everyday success habits

It's been said that a trainer teaches you how to do a task, and an educator teaches you a new way of thinking or seeing the world. My hope is that what I offer is a new way of thinking about how we can win client business – a conceptual framework with ample room for your own fingerprints. Rather than a sheet of music, my approach provides you with the instruments to create your own music – a song with its own unique tempo and melody.

References

What We Can Learn from the Beatles, Led Zeppelin, and the Rolling Stones story: Joan Magretta. *Understanding Michael Porter: The Essential Guide to Competition and Strategy*. Harvard Business Review Press, 2012.

From author's lecture notes, Senior Capstone Course on Business Strategy, BGEN499, Montana State University, 2017.

CHAPTER

25

Finding the Work That You Love

And What to Do When You Don't

You can only be truly accomplished at something you love. Don't make money the goal. Instead, pursue the things you love doing, and do them so well that people can't take their eyes off of you. All the other tangible rewards will come as a result.

—Maya Angelou, American poet

Francis Galton may not be as famous as his half-cousin Charles Darwin, but in the late 1800s he was a prolific English scholar, producing over 340 books and papers. He coined the term "nature versus nurture." It's hard to put Galton neatly in a box – his work interests ranged from sociology to psychology to anthropology – but among his most notable works is his study of high achievers.

In his book *Hereditary Genius*, published in 1869, Galton laid out the findings from his research of high achievers in science, athletics, music, poetry, and law. He concluded that all top performers demonstrated three distinctive qualities:

- Unusual ability
- Capacity for hard labor
- Exceptional zeal

Charles Darwin also believed in the importance of "zeal." In fact, after reading the first 50 pages of his half-cousin's book, he wrote, "I have always maintained that men do not differ much in intellect, only in zeal and hard work; and I still think these are the eminently important differences."

I can vividly remember the first time I heard "passion" used in business. I was in my early 20s and trying to figure out my career. In his keynote address at GE's 1991 annual shareholders meeting, Jack Welch described the traits that he looked for in his company's leaders. I don't remember his exact phrasing, but it was something along the lines of having a passion for your work. I shifted nervously in my seat, wondering, "Do I have passion for my work?" I wasn't sure I did. "Maybe I don't." I reflected. It was an unsettling feeling.

Do You Enjoy Your Work?

Is Galton's zeal the same as Jack Welch's passion? I think they are both getting at the same idea. Whether we call it zeal or passion, enjoying our work is an important component of success in any profession. Independent of innate skill or talent, it's hard to be really good at something that we don't enjoy. Whether a bronze sculptor, an NBA point guard, or a corporate trial attorney, we won't put in the time and effort to become an expert if we don't enjoy what we're doing.

In her book, *Grit: The Power of Passion and Perseverance*, Angela Duckworth tells the story of Chia-Jun Tsay. Chia holds several degrees from Harvard – her first being a bachelor's in psychology. She went on to earn two master's degrees in history and social psychology. And, as if she had too much time on her hands, while completing her PhD in organizational behavior she also picked up a second PhD in music as a concert pianist.

Chia went on to a successful academic career at University College London's School of Management. Along the way she performed musically at Carnegie Hall and the John F. Kennedy Center for the Performing Arts. When asked what drives her music while juggling the demands of her academic career, Chia says, "I loved music so much I practiced four to six hours a day throughout childhood." And, in college, she made time to practice almost as much. Yes, she had natural talent, but there was also a passion for music.

Clearly, Chia-Jun Tsay is what Francis Galton would call a high achiever. Most of us don't hold multiple PhDs from Harvard, and won't play a piano concerto for a U.S. president, in addition to teaching at a prestigious university. But the lessons of Chia's success apply to all of us. Whether we aspire to become a partner in a regional accounting firm or the co-founder of a digital advertising agency, we must find enjoyment in our work or we won't put in the time to become really good at what we do – in the words of Malcolm Gladwell, our 10,000 hours of practice.

If you haven't found your zeal or passion, you're in good company. In their *New York Times* best-seller, *Designing Your Life*, Bill Burnett and Dave Evans explain that only one in five young people under the age of 26 have a clear vision of what they want, what they want to accomplish in life, and why. And, from their research at Stanford, "80% of people of all ages don't really know what they are passionate about."

Enthusiasm for Our Work Is Key to Our Success

What role does passion play in the client's decision-making journey? I prefer the word "enthusiasm," because I believe the word "passion" – like zeal – has lost a lot of its original meaning. When narrowing down a short list of providers, I have come to believe that clients choose those who show enthusiasm for what they do.

Enthusiasm is contagious. Clients pick up on our love for what we do, and I believe it plays a substantive role in the decision process. Alternatively, when we are burned out or lack a real interest in our work, it affects those around us – colleagues, staff, and clients.

We must believe that what we do matters and is important. We must believe that we are truly helping our clients. When we don't have an enthusiasm for our work, clients will often choose someone who does.

To illustrate my point, take two equally talented athletes – say, two minor league baseball pitchers who were drafted at the same time. Both are equally talented and have the capability to make it to the big leagues. The first rookie loves the sport and enjoys practicing. He is a student of the culture and history of the sport. He loves the skill drills and the hours studying film on other great pitchers. The first rookie listens carefully to his coaches and tries to get better every day.

The second pitcher, equally talented, finds the work a bit tedious; it's like a job to him. He also trains hard, but he doesn't sit up late into the night studying film. He listens to his coaches but he doesn't ask nearly as many questions. All other things being equal, which player do you think makes it to the major league?

Apply this logic to any profession – raising growth capital or building a brand strategy. Or advising a client on a retirement plan or assisting in a patent filing. Two individuals with equal talent and equal qualifications; one has a love for their work and the other doesn't. Which one do you think will ultimately succeed?

Clients can sense the joy in our work; the love for what we do. We seem more interested in our work because we *are* more interested. You can't fake true love for what you do. Those who enjoy their work more often win the client's business.

Keep Exploring Until You Find Your Calling

It took me nearly 10 years of searching to find my professional footing. In my case, it wasn't so much that I understood that importance of passion; I was just driven to find work that I enjoyed. If I was going to spend decades of my life in earning a living, I wanted to do it pursuing something that I liked.

My own career journey led to the advice I give to my college seniors: explore while you're in your 20s and early 30s. It's hard to know what your passion will be until you find it. I've come to believe that we don't actually find our passion, but rather our passion finds us – if we're open to possibilities and willing to keep searching until we find it.

Our interests may change over the span of our career. Bill Burnett and Dave Evans, the Stanford professors we met earlier, offer some excellent career advice. They suggest that we remain curious, be willing to try stuff, ask for help when we need it, and realize it's a process.

If you don't love what you do today, it doesn't necessarily mean you need a complete career change. It could be that you simply need to look into other areas within your profession, perhaps a different niche within your field, or a new firm with a different mission.

An accountant friend of mine felt unfulfilled in auditing until she found her calling in serving nonprofits. She explained, "When I discovered the nonprofit world, my career began to blossom. I still practice accounting every day, but in serving nonprofits it feels totally different to me. I discovered that I do love accounting; I just wasn't interested in working on large corporate accounts."

Another friend was unhappy working as a Wall Street investment banker. He found his passion as a solo practitioner helping small business owners and entrepreneurs. He's still doing investment banking work every day, but, as he told me, "The work gets me up excited each morning. I'm making less than I made in New York, but I go to bed at night more satisfied with my work."

Respect, Trust, Likeability, and Enthusiasm

As we've explored the client's buying decision journey, we've discovered that clients hire someone they have come to know, respect, and trust, or who comes recommended by a colleague or friend. To this we later added the role of being likeable,

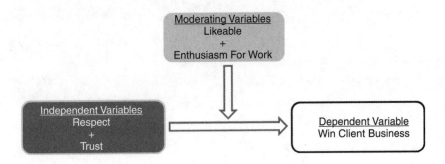

FIGURE 25.1 Being Likeable and Having Enthusiasm Allows Respect and Trust to Grow

in the context of being caring and interested in others. Being empathetic opens the gateways of respect and trust.

To these – respect, trust, and likeability – I'll add a fourth: enthusiasm for our work. (See Figure 25.1.) I have come to believe that clients choose professionals who enjoy their work. When we do, our enthusiasm is contagious. Between two equally qualified candidates, clients will choose the one who displays more enthusiasm for their work.

As humans, we are hard-wired to search for meaning. Work helps bring meaning and purpose to our lives. It allows us to feel useful and valuable; a contributor in making the world a better place. Finding work that brings us enjoyment is important to our personal happiness. And the happier we are with our work, the more successful we will be. I'm confident that you will find work that brings meaning and enjoyment to your life. And, when you do, it will positively impact your ability to win client business.

References

Francis Galton story: Martin Brookes. *Extreme Measures: The Dark Visions and Bright Ideas of Francis Galton*. Bloomsbury, 2014.

Chia-Jun Tsay story: Angela Duckworth. *Grit: The Power of Passion and Perseverance*. Scribner, 2016

Finding Your Passion story: Bill Burnett and Dave Evans. *Designing Your Life: How to Build a Well-Lived, Joyful Life*. Alfred A. Knopf, 2016.

Accountant story: Inspired by real characters known by the author. Names and locations have been changed.

CHAPTER

26

A High Road with a Long View

Parting Words as You Begin Your Rainmaker Journey

We do not receive wisdom, we must discover it for ourselves, after a journey through the wilderness which no one else can make for us, which no one can spare us, for our wisdom is the point of view from which we come at last to regard the world.
— Marcel Proust, French writer, *In Search of Lost Time*

I was listening to a *How I Built This* podcast episode recently where the host, Guy Raz, interviewed the Soul Cycle co-founders, Julie Rice and Elizabeth Cutler. During the interview, the co-founders shared a business mantra that guides their daily decision making: High road, long view. I rewound it a few times to hear it again. Their philosophy struck me as profound. And I have adopted it as a guiding principle in my own rainmaker journey.

When you're taking the high road with a long view, you need patience and persistence.

On Patience

There is a strong sense of urgency in American culture. Everything from Nike's *Just Do It* campaign to a George S. Patton's advice to his troops during WWII: "A good plan…executed now, is better than a perfect plan next week."

This is not inherently bad advice; it nudges us to action when we're caught up in indecision. Having a healthy bias for action is good, to the extent that we don't lose sight of the fact that our rainmaker journey will be long.

In an era of YouTube millionaires, it's easy to fall into the trap that success happens overnight – or at least by the close of business on Friday. If I were the genie in *Aladdin*, I'd grant you instant success. But I'm clearly not a genie and thus I must be honest with you: your success will not happen overnight.

We all remember the hit song *Who Let the Dogs Out*, but we've long forgotten the artist. The issue with instant stardom is that it can disappear as quickly as it arrived. The rainmaker's journey is not a get-rich-quick plan, although I'm sure if you Googled such a thing you'd find it. No, our journey will be long and we now have a road map for it.

Perhaps a better perspective is to see ourselves as the architect of the Great Pyramid of Egypt, building a lasting work of art one day at a time. You will lay one stone each day with purpose and a plan. On the last day of your career, you will lay the final capstone.

The benefits of patience sneak up on us like frequent flyer miles. The longer we are at our work, the easier it becomes to win client business. Our reputations will expand over time to the point where one day we are the obvious choice for our prospective clients. Our Rolodex will grow one relationship at a time until we are known and respected by those in our profession. Our resumes will accumulate experiences like the growth rings of a tall spruce until we have a right to a seat at the table.

I can remember how impatient I was when getting started. I wanted success now, or at least next month. I hungered for it. I easily succumbed to the belief that I could have success quickly. But I didn't. And most won't. And even if I could have had quicker success, I wouldn't go back and change a thing.

I've come to realize that fulfillment in one's career comes not from success, but rather the hard work that is required to achieve it. I'm clearly not alone in this belief, and certainly not the first to realize it. Benjamin Cardozo, the early twentieth-century Supreme Court justice, once wrote:

> In the end the great truth will have been learned, that the quest is greater than what is sought, the effort finer than the prize, or rather that the effort is the prize, the victory cheap and hollow were it not for the rigor of the game.

Enjoy each day realizing that you're doing important work. And have patience that in working your plan, you will achieve success in your career. And the success will be satisfying in knowing that you built it with much hard work and joy over many years. In the rainmaker's journey, patience is the fastest way to get what you want.

On Persistence

Persistence is the twin sidebar of patience on the ladder of success. While patience gives us the peace of mind in knowing we are headed in the right direction, persistence keeps us going in the face of setbacks. We will be knocked down in our pursuit of winning client business. Our journey will have unexpected challenges and detours, but you will achieve success with persistence.

Many cultures espouse the value of persistence, including Judeo-Christian, Buddhist, Islamic, and Hindu. But persistence is often misunderstood. Persistence is often portrayed as determination, rolling the rock uphill today only to find it rolling back down tomorrow. We are taught to believe that if we just keep persevering we'll eventually get to the top.

This is not the kind of persistence that I've come to believe in. Rather than persistence in doing the same things over and over that aren't working out, keep trying new approaches until you begin to gain traction. Or in the words of mathematician, James Yorke, "The most successful people are those who are good at Plan B."

Another take on persistence is to find what Alex Banayan calls the third door. In his book, *The Third Door: The Wild Quest to Uncover How the World's Most Successful People Launched Their Careers*, Banayan observes:

Life, business, success…it's just like a nightclub. There are always three ways in.

There's the First Door: the main entrance, where the line curves around the block; where 99% of people wait around, hoping to get in.

There's the Second Door: the VIP entrance, where the billionaires, celebrities, and the people born into it slip through.

But what no one tells you is there is always, always…the Third Door. It's the entrance where you have to jump out of line, run down the alley, bang on the door a hundred times, crack open the window, sneak through the kitchen – there's always a way.

Whether it's how Bill Gates sold his first piece of software or how Steven Spielberg became the most successful studio director in Hollywood history, they all took…the Third Door.

This is the kind of persistence I've come to admire. I've observed this trait in successful rainmakers in trying new things until one finds success. And when we find something that works for us in winning client business, we keep persisting until it doesn't work anymore. And then we move on to Plan B.

There's another aspect to persistence that rainmakers also have: the will to keep working at winning client business in the face of rejection. Winning client business may be the most maddening thing we'll ever practice in our careers. Comparatively, passing our professional certifications is easy, whether it's accounting, law, architecture, or engineering. Every successful rainmaker has countless disappointments; clients will suddenly shift gears at the eleventh hour to go with a competitor, and others will ghost us after months of productive discussion.

Give up on the notion of perfection in winning client business. It won't happen. Perfection – like infinity – is a theoretical construct that doesn't exist. Failure is a part of the process of being a rainmaker. We have to develop a capacity to keep swinging after we've struck out the previous three trips to the plate.

So, be persistent in tinkering with different approaches to winning client business. Experiment until you find what works for you. And have the persistence to overcome failure. We will fail. All rainmakers do. And, tomorrow, the sun will rise and we'll get up with renewed optimism to give it another go.

The Rainmaker Mindset

Here's what *high road, long view* means to me. Our journeys will be long – Tolkienesque adventures into the high mountains, not a weekend trip to the beach. It represents the rainmaker's belief in service not selling, character not charisma, and help not hype. We'll take the high road in doing what is right for our clients and business partners, knowing that in the end, our clients will trust us with more business and referrals.

For decades now, I've been a student of how clients buy and how to win client business. I readily admit that I geek out on these topics more than most. And, frankly, my enthusiasm for this doesn't offer many benefits at cocktail parties. That being said, after 25 years on this quest, I'm still passionate about the journey. While I know a lot more than I knew when I was getting started, this field is so deep and wide that I'm still learning new things every day.

I hope that my interest in the client's buying decision journey has inspired you. And I wish for you a professional career full of success and enjoyment.

References

George S. Patton quote: Martin Blumenson. *The Patton Papers: 1940–1945.* Houghton Mifflin, 1974.

Benjamin Cardozo quote: Andrew L. Kaufman. *Cardozo.* Harvard University Press, 1998.

Persistence story: Alex Banayan. *The Third Door: The Wild Quest to Uncover How the World's Most Successful People Launched Their Careers.* Currency, 2018.

High Road, Long View story: Guy Raz. "SoulCycle: Julie Rice & Elizabeth Cutler." *How I Built This* podcast, *January* 7, 2019.

Appendix A: The Rainmaker Skills Self-Test

What Are My Rainmaker Strengths and Weaknesses?

The Rainmaker Skills Self-Test is a diagnostic tool to gauge where you and your team are doing well, and where there may be opportunities for improvement. You may also have your colleagues rate how well you are doing in each of these areas as a check on your own perceptions. I hope this short questionnaire leads to honest self-reflection and interesting discussions among your teammates.

For each of the following statements, rate your level of agreement from 0 to 5 using the following scale:

5 = Strongly Agree

4 = Agree

3 = Somewhat Agree

2 = Somewhat Disagree

1 = Disagree

0 = Strongly Disagree

Skill 1: Create Your Personal Brand Identity

- I am clear about the topics for which I want to be known as the go-to expert.
- Others can easily explain what my go-to expertise is and who I wish to serve.
- I am recognized by my colleagues as a go-to expert in at least one area.
- I have a clear target audience of those I wish to serve.
- My target audience recognizes me as one of the top three for the services I provide.

213

Skill 2: Demonstrate Your Professional Expertise

- I recognize that prospective clients often struggle in their buying decision journey because it is difficult for them to gauge who the real experts are.
- I believe that demonstrating my professional expertise will strongly influence my ability to win new client business.
- I regularly invest time in providing my target audience with credibility markers that demonstrate my professional expertise.
- Others can easily see from my experiences, roles, activities, and credentials that I am an expert in my chosen field.
- It is clear from my personal online information that I am a leading expert in my field.

Skill 3: Build Your Professional Ecosystem

- I recognize that most of my success at winning client business will be through my professional ecosystem.
- I get to know people naturally through involvement in organizations, events, and activities that align with my interests.
- I invest the time in getting to know those in my professional ecosystem.
- I have segmented my professional ecosystem into groups according to their relative importance.
- I regularly stay in touch with the individuals in my professional ecosystem to remain top of mind.

Skill 4: Develop Trust-Based Relationships

- I recognize that trust is formed by showing others that we care for them and have their best interests at heart.
- I regularly do things for those in my professional ecosystem to show that I am thinking of them and want them to succeed.
- I believe that finding common ground with others is a proven way to build trust-based relationships.
- I readily share with prospective clients who I am, what I believe, and what my interests are.
- My colleagues would agree that I have a reputation for seeking win-win solutions.

Skill 5: Practice Everyday Success Habits

- I have developed a personal rainmaker approach that fits my strengths and preferences.
- I do not let daily urgencies distract me from practicing the rainmaker skills every day.

- Every day I connect with a few individuals in my professional ecosystem to show them that I care.
- I reflect regularly on my personal brand identity to make sure it is aligned with my aspirations.
- I regularly demonstrate my professional expertise through a commitment to my chosen credibility markers.

The Rainmaker's Journey

- I believe that successful rainmakers have achieved success through a committed effort to practicing the rainmaker skills.
- I do not try to be Wonder Woman or Superman, but focus instead on being the best version of myself that I can be.
- I enjoy my work and my clients can sense my enthusiasm for what I do.
- I have a long-term vision for what I want to achieve professionally.
- I believe in the rainmaker's mindset, where the focus is on service not selling, character not charisma, and help not hype.

After you have completed the assessment, average your responses for each question category to identify patterns. Where are you strong? What areas do you need to improve? You may also compare individual responses to team averages to facilitate discussion on best practices.

To benchmark your answers with others, go to www.fletcherandcompany.net/rainmaker.

Recommended Reading

Banayan, Alex. *The Third Door: The Wild Quest to Uncover How the World's Most Successful People Launched Their Careers*. Currency, 2018.

Burnett, Bill, and Dave Evans. *Designing Your Life: How to Build a Well-Lived, Joyful Life*. Alfred A. Knopf, 2016.

Cain, Susan. *Quiet: The Power of Introverts in a World That Can't Stop Talking*. Crown Publishers, 2012.

Cuddy, Amy. *Presence: Bringing Your Boldest Self to Your Biggest Challenges*. Little, Brown and Company, 2015.

Duckworth, Angela. *Grit: The Power of Passion and Perseverance*. Scribner, 2016.

Enns, Blair. *The Win Without Pitching Manifesto*. Rockbench Publishing, 2010.

Ferrazi, Keith. *Never Eat Alone: And Other Secrets to Success, One Relationship at a Time*. Currency/Doubleday, 2005.

Gensler, Art, with Michael Lindenmayer. *Art's Principles: 50 Years of Hard-Learned Lessons in Building a World-Class Professional Services Firm*. Wilson Lafferty, 2015.

Grant, Adam. *Give and Take: A Revolutionary Approach to Success*. Viking, 2013.

Hall, John. *Top of Mind: Use Content to Unleash Your Influence and Engage Those Who Matter to You*. McGraw-Hill, 2017.

Harding, Ford. *Creating Rainmakers: The Manager's Guide to Training Professionals to Attract New Clients*. John Wiley & Sons, 2006.

———. *Rain Making: Attract New Clients No Matter What Your Field*. Adams Business, 2008.

Kahneman, Daniel. *Thinking, Fast and Slow*. Farrar, Straus and Giroux, 2011.

Kleon, Austin. *Show Your Work! 10 Ways to Share Your Creativity and Get Discovered*, Workman Publishing Co., 2014.

Lewis, Michael. *The Undoing Project: A Friendship That Changed Our Minds*. W.W. Norton & Company, 2017.

Magretta, Joan. *Understanding Michael Porter: The Essential Guide to Competition and Strategy*. Harvard Business Review Press, 2012.

Maister, David. *Strategy and the Fat Smoker: Doing What's Obvious But Not Easy.* Spangle Press, 2008.

Miller, Donald. *Building a Story Brand: Clarify Your Message So Customers Will Listen.* HarperCollins, 2017.

Ries, Al. *Focus: The Future of Your Company Depends on It.* HarperBusiness, 1996.

Ries, Al. and Jack Trout. *Positioning: The Battle for Your Mind.* McGraw-Hill, 1981.

Thaler, Richard H., and Cass R. Sunstein. *Nudge: Improving Decisions About Health, Wealth, and Happiness.* Yale University Press, 2008.

Townsend, Heather, and Jon Baker. *The Go-To Expert: How to Grow Your Reputation, Differentiate Yourself from the Competition and Win New Business.* Pearson, 2014.

Vanderkam, Laura. *What the Most Successful Do Before Breakfast.* Portfolio/Penguin, 2012.

Acknowledgments

Every book is a team effort. It takes the talents of so many to see a book through to publication. I am deeply grateful for the caring support and guidance that many have provided me in this book's journey.

I'll start with my agent, Sheree Bykofsky. She believed in this book's potential before anyone else, and helped shape the original concept into what it has become. Thank you, Sheree, for your faith in me. And to my wonderful team at John Wiley & Sons, specifically Richard Narramore, Mike Campbell and Dawn Kilgore. Your early advice in the conceptual phase was filled with wisdom from many years in the book business. Mike, you were much more than an editor to me – you were a coach and mentor. Thank you also to Kelly Talbot and Amy Handy for polishing my work into a book ready for a public audience.

I am deeply indebted to Tom McMakin and his team at Profitable Ideas Exchange (PIE). Tom's support during all phases of this book was more than I deserved. Thank you to PIE's Jacob Parks and Carlie Auger for assistance with the rainmaker interviews that help bring these pages to life.

Thank you to the successful rainmakers whose personal stories are sprinkled throughout the book. Your advice is worth more than gold: Frank Bush, Graham Anthony, Judy Selby, Chuck McDonald, Eugene Buff, John Zombro, Eric Gregg, Murray Joslin, Charles Moren, Terry Pappy, Noel Sobelman, Michael Kelly, Mike McCracken, Scott Pollan, Cliff Farrah, Bill Stoddart, Doug Hall, Pete Sackleh, Walt Shill, Dominic Barton, Paddy Fleming, Tim Nath, Amir Tohid, Howard Hull, Keith Latson, Tim Hartland, Billy Newsome, and Michael Hinshaw.

Each writer is simply passing the torch on to the next generation. My ideas have been shaped by those who have come before me. I would like to especially thank David Maister and Ford Harding for sharing your knowledge with us through your many excellent books. I hope that some of your wisdom shines through in these pages and will carry on for many years to come.

While publishing a book is a team sport, writing a book is a deeply personal experience. One sits at a keyboard in solitude for hundreds of hours, often at night and on weekends. It's a long road filled with many hills and valleys. I couldn't have made it without the loving support of Bridget Maren Hoopes. I am also grateful for the generous support of my friends and family, and particularly my university colleague Ed Gamble – you are a true friend.

And, perhaps most of all, I would like to thank my readers. This book is for you. Each day I am inspired by the talented professionals who are on their own journeys to better understand how clients buy and how to win client business.

About the Author

Doug Fletcher is a leading expert on the topic of client development with 25 years of experience in management consulting. Doug helps professionals and organizations win more client business through a better understanding of the client's buying decision journey. He is the co-author of *How Clients Buy* (John Wiley & Sons, 2018).

Doug splits his time between speaking, writing, coaching, and teaching. In addition to his consulting practice, he serves on the board of directors of the Beacon Group, a growth strategy consulting firm headquartered in Portland, Maine.

From 1999 to 2014, Doug was co-founder and CEO of North Star Consulting Group, a technology-enabled consulting firm specializing in global web-survey projects. Earlier in his career, Doug was a management consultant with A.T. Kearney, and was trained at General Electric in its leadership development program. Doug earned an MBA from the Darden School of Business at the University of Virginia, and a BS in Electrical Engineering from Clemson University.

Doug lives in Montana and enjoys spending time in the outdoors with his friends and family.

He may be reached at: doug@fletcherandcompany.net.

www.fletcherandcompany.net

Index

Page numbers in *italics* reference figures.